SOLID STATE TRANSFORMATIONS

Library of Congress Catalog Card Number 66-18733

This Special Research Report is a translation of Sections 1 and 4 of *Mekhanizm i Kinetika Kristallizatsii*, edited by N. N. Sirota, F. K. Gorskii, and V. M. Varikash, published by Nauka i Tekhnika Press in Minsk in 1964 for the Crystal Production Section of the Scientific Council on Solid State Physics of the Academy of Sciences of the Belorussian SSR, Department of Solid State Physics and Semiconductors of the Academy of Sciences of the Belorussian SSR. The Russian text from which the translation was prepared was thoroughly corrected by the editors.

ISBN 978-1-4899-4754-3   ISBN 978-1-4899-4752-9 (eBook)
DOI 10.1007/978-1-4899-4752-9

SOLID STATE
TRANSFORMATIONS

Edited by
N. N. Sirota, F. K. Gorskii, and V. M. Varikash

Institute of Solid State Physics and Semiconductors
Academy of Sciences of the Belorussian SSR, Minsk

Translated from Russian by
Geoffrey D. Archard

Springer Science+Business Media, LLC

1966

This volume comprises the translation of Sections 1 and 4 of *Mechanizm i Kinetika Kristallizatsii (Mechanism and Kinetics of Crystallization)* edited by N. N. Sirota, F. K. Gorskii, and V. M. Varikash. The translation of Sections 2 and 3 is available from Consultants Bureau in a companion volume, under the same editorship, entitled *Crystallization Processes*.

FOREWORD

Detailed study of the growth of crystals from solutions and melts together with an examination of the processes of mass crystallization indicate that, for a quantitative description of the laws governing the generation and growth of crystals the earlier accepted mechanisms of the development of two-and three-dimensional nuclei are insufficient. Consideration must be given to amorphization of the growing crystal surfaces, the structure of the original phase, the possibility that the original phase may contain various complexes which participate in the formation of crystals, the presence of dislocations in the developing crystal, and the possibility of transformation without nuclei. Special significance attaches to the structural relationships of the original and developing phases, concentration fluctuations, and allowance for external actions, such as various fields and stresses. We find it thus necessary to consider a number of specific competing transformation mechanisms.

The need also exists for the further development and improvement in the field of phase transformation of the second kind. We can hardly fail to take account of the finer details of phase transformations, with special attention to changes in internal parameters as well as external equilibrium factors. In particular, the theory of phase transformations of the second kind which is based on the resolution of the thermodynamic potential into a power series of the internal parameter near the Curie temperature, cannot at present be considered altogether satisfactory, not only because of its limited nature and insufficiently founded approach to the analysis of the peculiarities of transformation, but also because it remains impossible to describe by means of this theory the various forms of transformations over a wide range of variations in temperature and pressure.

Among recrystallization processes, a unique position is occupied by phase transformations which take place in solids (polymorphic transformations, transformations of the martensite type, etc.).

This book considers some important problems of crystallization and phase transformations related to certain questions of particular importance in contemporary physics.

The first section, devoted to phase transformations, considers some general questions of the thermodynamics of critical phenomena. Semenchenko's article on the "Thermodynamics of Mesophases," in particular, serves as introduction to a large number of papers by the author and his colleagues on the thermodynamic analysis of certain features of transformations in the field of critical phenomena, based on Gibbs' conception of the thermodynamic instability of phases. Various papers in this section treat specific phase transformations of the second kind and associated phenomena of the scattering of ultrasound, X rays, and so forth.

Of great significance, in both theoretical scientific and applied areas, are transformation and recrystallization processes in the solid state. In the second section of the book articles are presented in which questions of polymorphism of the elements are considered in connection with their position in the periodic table, together with the laws of polymorphic transformations. In addition, several fine experiments on the growth of crystals in the solid phase during polymorphic transformations are described, the effect of vacancy diffusion on crystal growth during recrystallization is examined, and consideration is given to peculiarities of the martensite transformation.

Despite the great advances in the experimental and theoretical study of phase transformation processes, crystal growth, and mass crystallization, much additional theoretical and systematic experimental investigation is still necessary to fulfill our present national requirements. The development of modern technology has hastened the demand for the production of single crystals with controlled impurities, density, and character of dislocations, and has confronted us with the problem of phase transformations, which to a great extent determine the tempo of technical progress.

A mastering of the techniques of growing single crystals and obtaining materials with the required properties demands a considerable extension of those experimental and theoretical studies having a direct bearing upon the understanding of micro- and macroscopic features in the development of phase transformations. The present volume, which brings together a large amount of experimental and theoretical results, will surely serve to stimulate further study in the interesting and promising field of phase transformations. Most of the articles in this collection have been presented at various All-Union Conferences on the Theory of Crystallization, Thermodynamics, and the Kinetics of Phase Transformations held in the city of Minsk.

N. N. Sirota

CONTENTS

CONTENTS

PUBLISHER'S NOTE

The following Soviet journals cited in this book are available in cover-to-cover translations:

Russian title	English title	Publisher
Doklady Akademii Nauk SSSR	Soviet Physics—Doklady	American Institute of Physics
Fizika metallov i metallovedenie	Physics of Metals and Metallography	Acta Metallurgica
Fizika tverdogo tela	Soviet Physics—Solid State	American Institute of Physics
Izvestiya Akademii Nauk SSSR: Otdelenie khimicheskikh nauk	Bulletin of the Academy of Sciences of the USSR: Division of Chemical Science	Consultants Bureau
Kolloidnyi zhurnal	Colloid Journal	Consultants Bureau
Kristallografiya	Soviet Physics— Crystallography	American Institute of Physics
Pribory i tekhnika éksperimenta	Instruments and Experimental Techniques	Instrument Society of America
Uspekhi fizicheskikh nauk	Soviet Physics—Uspekhi	American Institute of Physics
Zhurnal fizicheskoi khimii	Russian Journal of Physical Chemistry	The Chemical Society (London)
Zhurnal neorganicheskoi khimii	Russian Journal of Inorganic Chemistry	The Chemical Society (London)

PART ONE

CRITICAL PHENOMENA AND PHASE TRANSFORMATIONS OF THE SECOND KIND

THERMODYNAMICS OF MESOPHASES

V. K. Semenchenko

The term "mesophase" was introduced by the first investigators of liquid crystals in order to emphasize their double character (fluidity as well as sharply expressed optical anisotropy). In the future we shall use this term to mean any phases which have properties intermediate between those of the phases bounding them, not in the sense of equilibrium coexistence, but rather as limiting states at the ends of some range of temperature and pressure. We shall show below that the experimentally observed existence of such phases has a thermodynamical explanation. Mesophases occur frequently and have a series of common properties which enable us to distinguish them from ordinary phases. The transformation of mesophases into ordinary phases takes place rather differently from phase transformations of the first kind, and, since it is not accompanied by the formation of a surface layer, it is simpler both from a thermodynamical and statistical point of view. The contradictions and difficulties arising in the theory of phase transformations are caused by ignoring the development of this surface layer, and we may therefore hope that it will prove possible to construct a strict, noncontradictory theory of mesophase transformations, and to pass from this to the theory of phase transformations of the first kind. Critical transformations constitute a limiting case of both transformations of the first kind and mesophase transformations.

The surface layer is an intermediate formation between two phases, not only in the goemetrical sense, but also with respect to properties, especially those characterizing the thermodynamic stability of the phases. Quantities characterizing stability are the derivatives of the thermodynamic forces T (temperature), p (pressure), and μ (chemical potential) with respect to their conjugate coordinates S (entropy), v (volume), C_j (concentration), and so forth. Let us call these quantities adiabatic coefficients of stability (ACS) or isodynamic coefficients of stability (ICS) [1]. Several experimental data [2] show that the coefficients of stability (CS) of the surface layer are smaller than in the volume phases. On raising T and P and approaching the critical point, the surface layer passes from monomolecular to polymolecular, and at the actual critical point absorbs both boundary phases, the whole system turning into a surface layer. Slightly above the critical point, however, phases with volumes below critical retain all the properties of the liquid phase, while phases with volumes above critical have all the properties of the gas phase, losing the characteristic precritical property of the coexistence of the two phases with surface tension at the surface separating them. On diagrams showing the variation of the derivatives of the generalized thermodynamic forces X_i with respect to the generalized coordinates as a function of the latter $(\partial X_i / \partial x_i) X_j - x_i$, they are now separated by a phase arising from the surface layer

and having intermediate properties — the mesophase. Transformation from liquid to vapor far from the critical temperature and pressure takes place through the surface layer. Above the critical point, this transformation must take place through the mesophase. This will be a hypercritical or mesophase transformation, which in contrast to the precritical and critical takes place over a certain range of temperature and pressure rather than at a single point T, P. The development of mesophases, however, takes place not only for the hypercritical transformation of liquid into vapor but also apparently for other states of aggregation, for example, a transformation from one crystalline form into another, or from the crystalline state into liquid (polymers). Under certain conditions, one of the bounding phases is in some cases not realized, and the substance exists only as mesophase and liquid.

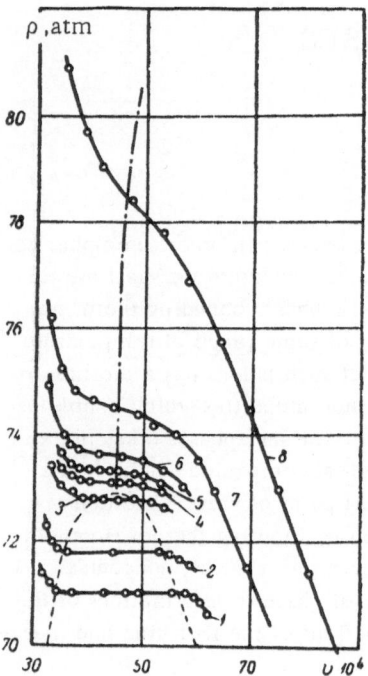

Fig. 1. CO_2 isotherms at various temperatures according to Michels: 1) 29.93°C; 2) 30.41; 3) 31.01; 4) 31.19; 5) 31.32; 6) 31.52; 7) 32.05; 8) 34.72°C (t_{cr} = 31.04°C; p_{cr} = 72.85 atm)

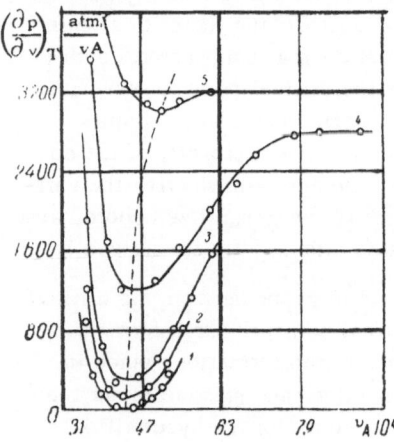

Fig. 2. $(\partial p/\partial v)_T$ −v diagram for CO_2 at various temperatures: 1) 31.01°C; 2) 31.52; 3) 32.05; 4) 34.72; 5) 40.09°C

Hypercritical State and Mesophases

Even in 1869, Andrews observed, on the basis of his own data, that there was a region with "intermediate conditions between the gaseous state and the volume characteristic of liquid," and concluded that "just as for gas below the critical temperature we have 1) the gaseous state, 2) transformation into liquid, and 3) the liquid state, so that for gas above the critical temperature we also have 1) the gaseous state, 2) an intermediate state corresponding to precipitation, and 3) conditions corresponding to a volume of liquid."* Andrews' statements were soon forgotten. The mere observations, however, of the p-v or T-S diagrams constructed from experimental data affords the realization (Fig. 1) that the facts are not as simple as they are presented in most books on thermodynamics and molecular physics. We see from Fig. 1 that there are points of inflection on the isotherms even above the critical one. It is known that the inflection points correspond to extrema on the curve of derivatives of the given function, i.e., in our case on the $(\partial T/\partial S)_p$ −S and −$(\partial p/\partial v)_T$ −v curves. Constructing the $(\partial p/\partial v)_T$ −v (Fig. 2) and $(\partial T/\partial S)_p$ −S (Fig. 3) diagrams, we see distinct extrema on these, while on the $(\partial p/\partial v)_T$ −v diagram for water (Fig. 4) we have both maxima and minima, indicating that on the hypercritical isotherms there is a second inflection point, the presence of which follows from the fact that, otherwise, the isotherms would cut the v-axis at finite values of v, i.e., the pressure of the gas would vanish for finite volumes and temperatures. We must emphasize that the extrema of all the

$$\left(\frac{\partial X_i}{\partial x_i}\right)_{X_j}$$

appear at the same T and p (see Fig. 5, which gives the reciprocal ICS. $(\partial S/\partial T)_p$, $(\partial v/\partial p)_T$, $(\partial v/\partial T)_p$]. But

$$\left(\frac{\partial X_i}{\partial x_i}\right)_{X_j}$$

characterizes the thermodynamic stability of the system, so that we may say that the region between the maximum and minimum of stability (shown by the dotted line in Fig. 4) constitutes a region of reduced stability of the system. If we consider the v-T diagram (Fig. 6), we see that in the precritical region the phase transformation is accompanied by a volume jump, shown as the dotted lines in Fig. 6. On the hypercritical curves, instead of a jump we have a rapid rise, gradually smoothing out as T and p increase. This rise corresponds to passage through the region of reduced stability, and the jump to a leap through a completely unstable region. Just as in the p-v and T-S diagrams in the precritical region in which we have a part bounded by a spinodal where the system is unstable with a vertex at the critical point, so also in the hypercritical region the $(\partial p/\partial v)_T$ −v diagram (see Fig. 4) has a region of reduced stability. We have

*Quoted from the Russian version of T. Andrews' "Continuity of the Gaseous and Liquid States of Matter" (GTTI, Moscow, 1933). K. V. Arkhangel'skii informed me of Andrews' observations.

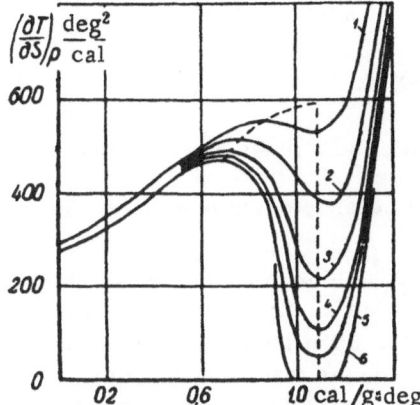

Fig. 3. The $(\partial T/\partial S)_p - S$ diagram for water, from the data of D. L. Timrot and A. E. Sheindlin, for various pressures (kg/ cm^2): 1) 900; 2) 600; 3) 400; 4) 300; 5) 240; 6) 200.

called the curve bounding this region the quasispinodal, and its highest point the supercritical point [4]. Figure 5 shows that not only all the CS but also derivatives with respect to nonconjugate variables, such as $(\partial T/\partial v)_p$, pass simultaneously through extrema in the hypercritical region (Fig. 5 shows the reciprocals of the CS). Hence the existence of a region of reduced thermodynamic stability is an experimental fact, established in the absence of any explanatory theory, although in order to establish this fact irrefutably, long-known experimental data had to be analyzed from a theoretical point of view. Such data were known as "anomalies" of the various substances.

Thermodynamics of Hypercritical Transformations

Let us pass on to the thermodynamic treatment of the facts stated. The system is in a state of thermodynamic equilibrium if its internal energy U is less than the energy of all neighboring states, i.e., the second differential or variation of U is positive:

$$\delta^2 U = \sum_i \sum_j \frac{\partial^2 U}{\partial x_i \partial x_j} \cdot \delta x_i \delta x_j > 0. \qquad (1)$$

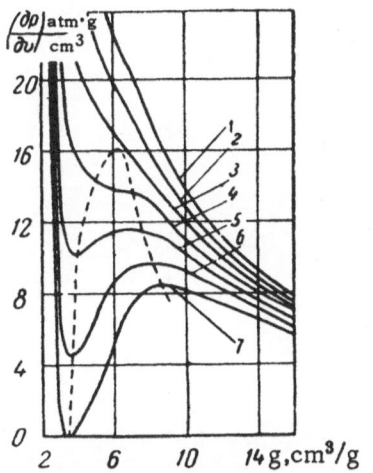

Fig. 4. The $(\partial p/\partial v)_T - v$ diagram for water, from the data of D. L. Timrot and A. E. Sheindlin, for various temperatures: 1) 430° C; 2) 420; 3) 410; 4) 400; 5) 390; 6) 380; 7) 370°C

The sum on the right in (1) is a quadratic form, positive if the determinant composed of the coefficients of $\delta x_i \delta x_j$ and its principal minors are positive. We therefore write the condition for the stability of thermodynamic equilibrium as

$$D_n = \begin{vmatrix} U_{11}, U_{12} \cdots U_{1n} \\ U_{21}, U_{12} \cdots U_{2n} \\ \cdot \quad \cdot \quad \cdot \quad \cdot \quad \cdot \\ U_{n1}, U_{n2} \cdots U_{nn} \end{vmatrix} > 0; D_{n-1} = \begin{vmatrix} U_{11}, U_{12} \cdots U_{1, n-1} \\ U_{21}, \cdots U_{2, n-1} \\ \cdot \quad \cdot \quad \cdot \quad \cdot \\ U_{n-1} \cdots U_{n-1, n-1} \end{vmatrix} > 0,$$

$$D_1 = U_{11} > 0; \ U_{ii} = \frac{\partial^2 U}{\partial x_i \partial x_j}. \qquad (2)$$

$$x_i = S, v, C_j \ldots : \ X_i = T, p, \mu_j \cdots$$

But $\partial^2 U/\partial x_i \partial x_j = \partial X_i/\partial x_j$, so that we may write (2) in the following way:

$$D_n = \frac{D(X_1, X_2 \cdots X_n)}{D(x_1, x_2 \cdots x_n)}, \qquad (3)$$

Since the choice of indices is arbitrary, all the $\left(\frac{\partial X_i}{\partial x_i}\right)_{x_j} > 0$ [this may also be shown from the inequalities in (2)]; we call $\left(\frac{\partial X_i}{\partial x_i}\right)_{x_j}$ the adiabatic or isocoordinate coefficients of stability (ACS) and D_n the

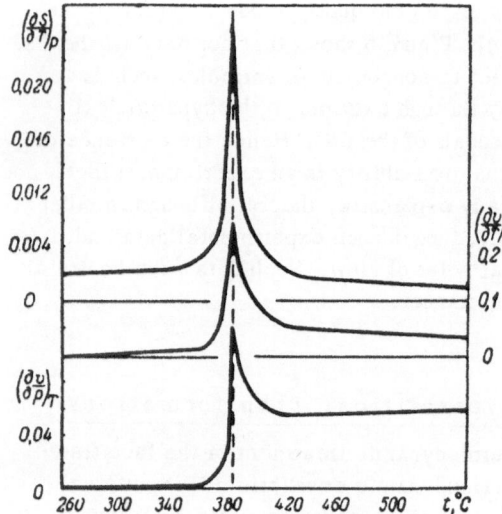

Fig. 5. Reciprocals of ICS $(\partial S/\partial T)_p$ $-(\partial v/\partial p)_T$ and $(\partial v/\partial T)_p$ for water, from the data of D. L. Timrot and A. E. Sheindlin, for pressure 260 kg/cm².

stability determinant. By the definition of a positive quality,

$$0 \leqslant D_n \leqslant + \infty \qquad (4)$$

As the theory of quadratic forms shows, however, the coefficients $\left(\dfrac{\partial X_i}{\partial x_i}\right)_{x_j}$ cannot equal zero, but must have nonzero positive values; hence, in stable equilibrium, all the ACS are nonzero and positive even for the case in which D_n reaches its lower limit ($D_n = 0$). Experimental determination of the ACS is sometimes troublesome, so that we may use the isodynamic CS obtained from the ACS by the transformation.

$$d x_1 = \sum_{i=1}^{n} \left(\frac{\partial X_i}{\partial x_i}\right)_{x_j} dx_i;$$

$$dX_n = \sum_{i=1}^{n} \left(\frac{\partial x_n}{\partial x_i}\right)_{x_j} dx_i, \qquad (5)$$

$$D'_n = \begin{vmatrix} X_{11} \cdots X_{1n} \\ \cdot \ \ \cdot \ \ \cdot \ \ \cdot \\ \cdot \ \ \cdot \ \ \cdot \ \ \cdot \\ X_{n1} \cdots X_{nn} \end{vmatrix} \cdots D_{xi} = \begin{vmatrix} X_{11} \cdots dX \cdots X_{1n} \\ \cdot \ \ \cdot \ \ \cdot \ \ \cdot \ \ \cdot \ \ \cdot \ \ \cdot \\ X_{n1} \cdots dX_n \cdots X_{nn} \end{vmatrix}, \qquad (6)$$

$$dx_i = \frac{DX_i}{D'_n} \qquad (7)$$

$$X_{ij} = \left(\frac{\partial X_i}{\partial x_j}\right)_{x_k}$$

Expanding the determinant in the numerator of (7) with respect to dX_i, we obtain

$$dx_i = \frac{1}{D'_n} \sum_i D_{ek} \, dX_i \qquad (8)$$

$$e, k \neq i; \ \left(\frac{\partial X_i}{\partial x_i}\right)_{x_j} = \frac{D_{ek}}{D'_n}. \qquad (9)$$

Let us take a simple example:

$$dT = \left(\frac{\partial T}{\partial S}\right)_v dS + \left(\frac{\partial T}{\partial v}\right)_s dv;$$

$$-dp = -\left(\frac{\partial p}{\partial S}\right)_v dS - \left(\frac{\partial p}{\partial v}\right)_s dv; \qquad (10)$$

$$dS = \frac{D_s}{D}; \quad dv = \frac{D_v}{D}; \qquad (11)$$

$$D = \begin{vmatrix} T_s, T_v \\ -p_s, -p_v \end{vmatrix}; \qquad (12)$$

$$D_s = \begin{vmatrix} dT, T_v \\ dp, -p_v \end{vmatrix}; \quad D_v = \begin{vmatrix} T_s, dT \\ -p_s, dp \end{vmatrix}; \qquad (13)$$

$$dS = \frac{-p_v dT - T_v dp}{D}; \quad dv = \frac{T_s dp + p_s dT}{D};$$

$$\left(\frac{\partial T}{\partial S}\right)_p = -\frac{D}{p_v}; \left(\frac{\partial p}{\partial S}\right) = -\frac{D}{T_v}; \left(\frac{\partial T}{\partial v}\right)_p = \frac{D}{p_s}; \left(\frac{\partial p}{\partial v}\right)_T = \frac{D}{T_s} \qquad (14)$$

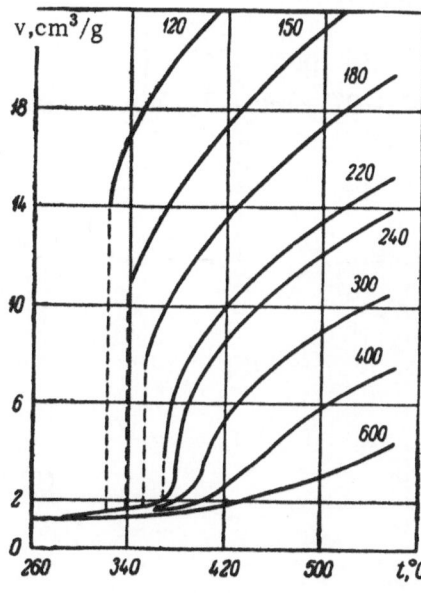

Fig. 6. Form of volume change as a function of temperature in the pre-critical and hypercritical regions.

Let us consider the properties of the system as it passes through a stability extremum. Since we assume that the stability is determined by the values of D, the stability extrema will be determined by the extrema of D:

$$dD = \sum_1^n \left(\frac{\partial D}{\partial x_i}\right)_{x_j} dx_i = 0; \left(\frac{\partial D}{\partial x_i}\right)_{x_j} = 0. \qquad (15)$$

For a simple system (x_i = S, v; X_i = T, p) we have

$$\left(X_{ix_i, x_j} = \frac{\partial^2 X_i}{\partial x_j \partial x_j}\right);$$

$$D = \begin{vmatrix} T_s, T_v \\ -p_s, -p_v \end{vmatrix}; \left(\frac{\partial D}{\partial S}\right)_v = -T_{ss} p_v - T_s p_{vs} - 2T_v T_{vs} = 0;$$

$$(T_v = -p_s); \left(\frac{\partial D}{\partial v}\right)_s = -T_{sv} p_v - T_s p_{vv} - 2T_v T_{vv} = 0. \qquad (16)$$

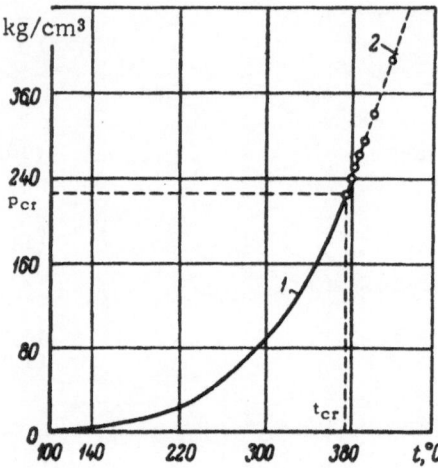

Fig. 7. Continuation of the p−T phase equilibrium curve into the hypercritical region.

The system of equations (16) is satisfied for

$$T_{ss} = p_{vv} = p_{vs} = T_{vs} = T_{vv} = 0, \qquad (17)$$

i.e., if all the adiabatic CS pass through an extremum. We shall show that if the ACS pass through extrema then the conjugate ICS and the derivatives with respect to nonconjugate coordinates also pass through extrema. Differentiating with respect to S and v (14) and taking account of (15) and (17), we obtain

$$\left(\frac{\partial^2 T}{\partial S^2}\right)_v = -\frac{1}{p_v}\left(\frac{\partial D}{\partial S}\right)_v + \frac{D}{p_v^2}p_{vs} = 0;$$

$$\left(\frac{\partial^2 p}{\partial v^2}\right)_s = \frac{1}{T_s}\left(\frac{\partial D}{\partial v}\right)_v - \frac{D}{T_s^2}T_{vs} = 0, \qquad (18)$$

whence we determine the geometrical locus of the external points on the $(\partial T/\partial S)_p$−S and $(\partial p/\partial v)_T$−v diagrams, i.e., the curves which we have called quasispinodals (see Figs. 2−4). The line of phase equilibrium, for example, the saturated vapor pressure line, can also be represented on the p−T diagram. Can the quasispinodal be projected on the p−T plane? Let us transform the differential equations for dp/dT to variables v, s from 10:

$$\frac{dp}{dT} = \frac{p_s ds + p_v dv}{T_s ds + T_v dv} = f(s,v). \qquad (19)$$

The right hand side of (19) is a function of s, v; let us find it complete differential:

$$df = \frac{p_{ss}ds^2 + 2p_{vs}dvds + p_{vv}dv^2}{T_s ds + T_v dv} - \frac{p_s ds + p_v dv}{(T_s ds + T_v dv)^2}(T_{ss}ds^2 + 2T_{sv}dsdv + T_{vv}dv^2). \qquad (20)$$

Substituting the values of p_{ss}, p_{vs}, p_{vv}, T_{ss}, T_{sv} from (17), we find

$$df = 0.$$

Hence, integrating, we obtain

$$f = \text{const} = \frac{dp}{dT}; \quad p - p_c = A(T - T_c). \qquad (21)$$

The quasispinodal begins at the critical and ends at the supercritical point, so that we obtain for A

$$A = \frac{p_{sc} - p_c}{T_{sc} - T_c}; \qquad (22)$$

$$p = p_c + \frac{p_{sc} - p_c}{T_{sc} - T_1}\,(T - T_c).$$

(23)

All the formulas obtained (15) to (23) have a general character, and can be written for any forces and coordinates X_i, x_i. Figure 7 shows that for hypercritical liquid—vapor transformations, (23) is well satisfied.

Let us now examine the molecular-statistical significance of the CS [7]. Twice differentiating the expression for the statistical analog of the Gibbs free energy F:

$$F = - kT\ln \int e^{-\frac{H}{kT}} d\omega \; ;$$

$$H = \frac{1}{2m}\sum p_i^2 + U\,(q);$$

$$\left(\frac{\partial^2 F}{\partial T^2}\right)_v = -\frac{C_v}{T} = \frac{\overline{H^2} - \overline{H}^2}{kT^3} = \frac{\overline{(H - \overline{H})^2}}{kT^3},$$

(24)

i.e., the reciprocal of the thermal ACS is proportional to the energy fluctuation. Analogous relations may be derived for other CS as well as for the mixed derivative in the stability determinant. This result makes the physical significance of the CS extremely clear (the greater the deviation of the system from the mean values, the easier it is to convert it into another state and the less stable its present state).

Let us briefly summarize the main results of our analysis.

1. The overall characteristic of any equilibrium state demands analysis of its stability, determined by the magnitude of the stability determinant D_n (2) and the ACS. The ACS can be transformed into ICS by (9) if the latter can be determined more easily experimentally.

2. The construction of the $\left(\frac{\partial X_i}{\partial x_i}\right)_{X_j} - x_i$ diagram enables us to form an idea of the variation of stability.

The lines giving the extrema of stability (quasispinodals) are given on the $(\partial X_i/\partial x_i)_{X_j} - x_i$ diagram (18). These lines can also be projected on the $P_i - T_i$ diagram (or on any $X_i - x_i$ plane), where they degenerate into a straight line connecting the critical and supercritical points.

3. From the molecular statistical point of view, the CS are quantities inversely proportional to the fluctuation, so that stability drops as fluctuations develop in the system.

Let us trace the path of a hypercritical transformation. Normally stability decreases as its conjugate coordinate increases (for example, v in Fig. 4). In the case of a hypercritical transformation, this decrease stops at a certain point, at which the stability passes through a minimum and fluctuations reach their greatest development; afterwards the stability again begins to rise. Individual regions of the system in the neighborhood of this point of least stability approach the bounding phases in respect to properties characterized by average statistics, taken only over the microregion in question, so that here we may say that the system has passed into a microheterogeneous state. These fluctuation "seed" phases, however, differ from macrophases in that their lifetime is short compared with the equilibrium lifetime of macrophases, and their dimensions are of the order of the thickness of the macrophase surface layer. Hence surface tensions at their boundaries, and indeed the boundaries themselves, do not exist. The concepts, discussed by some investigators, of "liquid—vapor mixtures," "solutions or emulsions of two phases," and so on, may well be visually clearer, but they are extremely coarse approximations, robbing the mesophase of its main properties, namely macroscopic homogeneity, and hence also the absence of any internal surfaces of separation and surface tension characterizing these. The increase in stability after the mesophase has passed through the stability minimum results in decay of the fluctuations resembling the low-temperature phase in their properties; the stability reaches a maximum corresponding to the phase "purest" with respect to microheterogeneity, and then falls monotonically. On the $(\partial T/\partial S)_p$-S diagrams, the maxima correspond to the smallest S (see Fig. 3). We will not analyze the relation between the

various $\left(\dfrac{\partial X_i}{\partial x_i}\right)_{X_j} - x_i$ diagrams, and will rather confine ourselves to the statement that the most straightforward and correct picture of the behavior of stability in the case of simple systems comes from the stability determinant D_n, set out as a D_n,T, p surface, or its projections on the D_n—T and D_n—p planes. We note further that, knowing the ICS, we can use their reciprocals $(\partial x_i/\partial x_j)_{X_k}$ to construct an "instability determinant" D^n, the reciprocal of which, from the theory of determinants, is D_n, as we can easily check from formula (14):

$$\frac{D(x_{11}x_2\cdots x_n)}{D(X_{11}X_2\cdots X_n)} = \frac{1}{\dfrac{D(X_1,X_2\cdots X_n)}{D(x_1,x_2\cdots x_n)}} = $$

$$= \begin{vmatrix} \left(\dfrac{\partial x_1}{\partial X_1}\right)_{x_i} \cdots \left(\dfrac{\partial x_1}{\partial X_n}\right)_{x_i} \\ \cdots \cdots \cdots \\ \left(\dfrac{\partial x_n}{\partial X_1}\right)_{x_i} \cdots \left(\dfrac{\partial x_n}{\partial X_n}\right)_{x_i} \end{vmatrix} = D^n. \tag{25}$$

The properties and behavior of D^n are the inverse of the corresponding properties of D_n: when D_n reaches the lower bound of stability, D^n tends to $+\infty$, and, on the other hand, as D_n approaches its upper bound of ∞, D^n approaches zero. Since the thermodynamic potentials cannot be purely forces (this follows from the Gibbs formula), D^n is not connected directly with any of the thermodynamic potentials.

Ferromagnetic and Ferroelectric Transformations as Mesophases

Ferromagnetic transformations are considered at the present time (as are many of the transformations in ferroelectrics) as a typical example of phase transformations of the second kind [8]. Although phase transformations of the second kind are often understood to mean any transformations which are not of the first kind, we shall consider Ehrenfest's [9] definition to be the closest to reality, this being expressed as

$$Z' \equiv Z''; \tag{26}$$

$$\left(\frac{\partial Z}{\partial X_i}\right)' X_i = \left(\frac{\partial Z}{\partial X_i}\right)'' X_i; \tag{27}$$

$$\left(\frac{\partial^2 Z}{\partial X_i^2}\right)'_{X_j} \neq \left(\frac{\partial^2 z}{\partial X_i^2}\right)''_{X_j}, \qquad (28)$$

in which (26) is superfluous, as follows from (27). According to (28), phase transformations of the second kind are characterized by jumps in the second derivatives $\left(\frac{\partial^2 Z}{\partial X_i^2}\right)_{X_j}$ of the potential constituting the force potential for single-component systems with constant mass. Consideration of existing experimental material, however, shows that for ferromagnetic and ferroelectric transformations no jumps in the second derivatives $\partial^2 U/\partial x_i \partial x_j$ are observed. Experiment shows the presence of maxima, in many cases extremely drawn out and feeble in relative magnitude. The confusion is still further complicated by the fact that the principal indication of a phase transformation of the second kind is considered to be the existence of a "jump" in specific heat, which is neither a CS nor the reciprocal of one, and all conclusions regarding the character of the phase transformations are drawn from the form of the $C_p - T$ curves. Moreover, the thermodynamic properties of ferromagnetics have up till now been poorly and unsystematically studied, though things are rather better for ferroelectrics. The traditional study of magnetism treats the Curie–Weiss law and the concept of the internal field as absolute. Statistical considerations indicate that the internal field is not a statistical mean entering into the statistical analogs of induction and dielectric and magnetic permeability, so that it cannot therefore be estimated from the values of these quantities.

These considerations give us a basis for regarding ferromagnetic transformations, and those ferrelectric transformations which are not of the first kind, as hypercritical, mesophase transformations. In this case, we may expect all the ACS and ICS to pass through minima, and the transformation points, i.e., the temperatures at which the CS pass through a minimum, to be displaced under the action of fields and stresses. This displacement should be a linear function of the mechanical and magnetic field intensities. No jumps should be observed in the CS. The best check on all this would appear to be a calculation of the stability determinant D and analysis of its behavior near the transformation point, but available experimental data will not allow this to be done. In order to calculate D, we must have a selection or all of the adiabatics $\partial^2 U/\partial x_i \partial x_j$ or isodynamics measured for a single-domain single crystal. As yet there are no such data, so that we must confine ourselves to examining the behavior of individual CS.

The behavior of a dielectric or magnetic is determined by three pairs of variables T, S, σ_{ij}, ε_{ij}, E, D, H, B, of which T and S are scalars, E, D, H, and B vectors, and σ_{ij}, ε_{ij} symmetric tensors. From these we can construct eight different thermodynamic potentials (if the system is single-component). The mesophase transformations are those for which the heat and all the works of transformation are zero, and all the CS, both adiabatic $\left(\frac{\partial X_i}{\partial x_i}\right)_{X_j}$, and isodynamic, pass through a minimum at a transformation point (which is in fact not a point, since the transformation itself consists of a transistion through a region of reduced stability) just as, according to (18), do all the isodynamic derivatives with respect to nonconjugated coordinates. Hence we have minima for

$$\left(\frac{\partial T}{\partial S}\right)_{V,D}, \quad \left(\frac{\partial E}{\partial D}\right)_{S,V}, \quad \left(\frac{\partial \sigma_{ij}}{\partial \varepsilon_{ij}}\right)_{S,D}; \qquad (29)$$

$$\left(\frac{\partial T}{\partial S}\right)_{p,E}, \quad \left(\frac{\partial E_i}{\partial D}\right)_{p,T}, \quad \left(\frac{\partial \sigma_{ij}}{\partial \varepsilon_{ij}}\right)_{T,x_{ij}} \left(\frac{\partial T}{\partial v}\right)_{E,X_{ij}}, \quad \left(\frac{\partial T}{\partial D}\right)_{E,X_{ij}};$$

$$\left(\frac{\partial E}{\partial S}\right)_{x_{ij},\,T}, \quad \left(\frac{\partial E_i}{\partial v}\right)_{p,T}, \quad \left(\frac{\partial \sigma_{ij}}{\partial S}\right)_{E,T}, \quad \left(\frac{\partial \sigma_{ij}}{\partial D_i}\right)_{E,T}$$

where σ_{ij} denotes any of the stresses σ, and the electric parameters E and D may be replaced by magnetic ones, if one is dealing with a magnetic material. Experimental data [8–10] show that for ferromagnetic transformations $c_{p,H}$ passes through a weak and rather extended maximum, and hence $T/c_{p,H}$ passes through a still less sharp minimum, μ and σ through a sharp maximum, and the magnetostriction $(\partial v/\partial E)_{p,T}$ through a maximum. Some of the derivatives in (29) have apparently not been measured at all. We shall not reproduce the corresponding graphs, which can be found in any book on ferromagnetism [8, 11]. Quite a lot of these are in a book by K. P. Belov, who, presenting a large number of curves with (for the most part) smooth extrema, assures the reader of the existence of jumps. Unfortunately, available experimental data is so unsystematic, and the data for different CS obtained for completely different conditions and different samples, that at present it is extremely hard to calculate reliable values of the stability determinant. The behavior of the stability determinant would doubtlessly give the most convincing proof that ferromagnetic transformations are of the mesophase type. A question long discussed by students of magnetism is the presence of residual magnetism after the Curie point; this is one of the clearest indications that we are here in fact dealing with a mesophase, and that the Curie point is simply the point of least stability. The error of the magnetologists lies in the fact that they consider the magnetic permeability as a property physically as valid as volume, entropy, or pressure, whereas the physical nature of permeability differs fundamentally from these: the former characterize mean equilibrium first derivatives of thermodynamic potentials, while μ is the second derivative of fluctuations, the mean square deviation from mean values. The growth and slow decay of fluctuations (not in time, but as one moves away from their maximum, the Curie point) is the principal sign of mesophase and critical transformations, including ferromagnetic.

Ferroelectrics behave in most cases analogously to ferromagnetics: derivatives (3, 4) pass through minima, and their reciprocals through maxima. The calculation of the stability determinant presents the same difficulties, since not a single investigator has measured the complete set of derivatives (3, 4) on a single crystal, or even on polycrystalline specimens of common origin. The existence of spontaneous magnetization does not emanate from the ordinary model of a ferromagnetic or dielectric (Van Leeven-Terletskii theorem), and is therefore considered as a quantum effect, though one can scarcely think that the properties of electrets are produced by any quantum phenomena. Since the charge carriers in ferroelectrics are ions, changes in their disposition may be associated with a change in the lattice type and may thus have any character, depending on the values of the external forces. In view of this, a study of the behavior of the CS, and especially the stability determinant, takes on a decisive significance in elucidating the type of phase transformation.

We shall now show that the molecular field does not enter into the expressions for induction and dielectric constant obtained by the statistical method in the form of a statistical mean, having the sense of a certain macroscopic quantity. Let us choose the function Φ as potential:

$$\Phi = U - TS + \frac{ED}{8\pi}\, v. \tag{30}$$

The statistical analog of this is

$$\Phi = - KT \ln \int e^{- \frac{E^2 v}{8\pi KT} - \frac{\Sigma p_i^2}{2mKT} - \frac{E}{KT} \mu \sum \cos \vartheta_i - \frac{\mu^2}{KT} \sum \sum \frac{f(\vartheta, \varphi)}{r^3}} d\mu \qquad (31)$$

Here, in the expression for the Hamiltonian we have introduced the field energy $(E^2/8\pi)$ v, into the exponent of the distribution function; this is a constant with respect to integration variables q and p, but depends on the field and the volume v of the system. Differentiating (31) with respect to E, we obtain the induction

$$\left(\frac{\partial \Phi}{\partial E} \right)_{T, v} = \frac{Dv}{4\pi} = \frac{Ev}{4\pi} + \frac{\int \mu \sum_i \cos \vartheta_i e^{-\frac{H_q}{KT}} d\omega_q}{\int e^{-\frac{H_q}{KT}} d\omega_q} ; \qquad (32)$$

$$H_q = \mu \sum \cos \vartheta_i + \frac{\mu^2}{2} \sum \sum \frac{f(\vartheta, \varphi)}{r^3} ; \quad D = E(1 + 4\pi \mu \sum \cos \vartheta_i). \qquad (33)$$

Differentiating again with respect to E, we obtain the dielectric constant ε:

$$\left(\frac{\partial D}{\partial E} \right)_{T, v} = \varepsilon_{T, v} = 1 + 4\pi \left\{ \frac{\int (\Sigma \mu \cos \vartheta_i)^2 e^{-\frac{H_q}{KT}} d\omega_q}{KT \int e^{-\frac{H_q}{KT}} d\omega_q} - \right.$$

$$\left. - \frac{1}{KT} \left(\frac{\int (\Sigma \mu \cos \vartheta_i) e^{-\frac{H_q}{KT}}}{\int e^{-\frac{T_q}{KT}} d\omega q} \right)^2 \right\} = \qquad (34)$$

$$= 1 + 4\pi \frac{\overline{(\Sigma \mu \cos \vartheta_i)^2} - \overline{(\Sigma \mu \cos \vartheta_i)^2}}{KT} .$$

Expressions (33) and (34) show that the values of the macroscopic parameters D and ε are determined by the mean values of ϑ_i, where ϑ_i is the angle between the direction of the field and dipole. The internal field of the dipole, the energy of which is given by the term $\mu^2 \Sigma [\Sigma f(\vartheta, \varphi) / 2r^3]$, affects the Hamiltonian inasmuch as ϑ also comes into it, thus depending both on the interaction of the dipole with other dipoles and on its interaction with the external field E. The mean value of the field acting on one dipole may be obtained by differentiating (31) with respect to μ:

$$\overline{E} = \left(\frac{\partial \Phi}{\partial \mu} \right)_{T, v, E} = \frac{\int \left\{ \mu \Sigma \Sigma \frac{f(\vartheta, \varphi)}{r^3} + E \Sigma \cos \vartheta_i \right\} e^{-\frac{H_q}{KT}} d\omega_q}{\int e^{-\frac{H_q}{KT}} d\omega_q} = \mu \Sigma \Sigma \overline{\frac{f(\vartheta, \varphi)}{r^3}} - E \overline{\Sigma \cos \vartheta_i}. \quad (35)$$

The difference between the total field $\bar{E}$ and the external gives us the internal field $\bar{E}_B$:

$$\overline{E} - \overline{E \, \Sigma \cos \vartheta_i} = \mu \, \overline{\Sigma \, \Sigma \frac{f(\vartheta, \varphi)}{r^3}}. \tag{36}$$

Statistical theory, as we see, offers the possibility of calculating all the fundamental characteristics of the dielectric (or magnetic, if we replace the field and dipole moments by the corresponding magnetic parameters).

Here we have selected mesophase transformations in the liquid—gas, magnetic, and dielectric systems. They are in fact found much more often. One of the typical mesophase transformations is the $\alpha-\beta$ transformation of quartz [12] and transformations in crystal polymers [13, 14]. In some cases one of the bounding phases may not exist under ordinary conditions, in which case the mesophase plays the dominant part. Protoplasm also apparently exists in a mesophase state, so that a study of the properties of mesophases and mesophase transformations has great significance for biophysics [15].

Literature Cited

1. V. K. Semenchenko, Zh. Fiz. Khim., 33:1440 (1959).
2. G. I. Berezin, A. V. Kiselev, and A. V. Sinitsyn, Doklady Akad. Nauk, 135:638 (1960); Kolloidnyi zhurnal, 23:638 (1961).
3. A. Michels, B. Blaise, and C. Michels, Proc Roy Soc. A 160:346 (1937).
4. V. K. Semenchenko, Dokl. Akad. Nauk 92:625 (1953); 99:1045 (1954); Collection: "Application of Ultra-acoustics to the Study of Matter" Vol. 3, p.51 (MOPI, Moscow, 1956).
5. V. K. Semenchenko, Selected Chapters on Theoretical Physics (GUPI, Moscow, 1960).
6. V. K. Semenchenko, Zh. Fiz. Khim., 36:1115 (1962).
7. V. K. Semenchenko, Zh. Fiz. Khim., 21:1461 (1947).
8. S. V. Vonsovskii and Ya. S. Shur, Ferromagnetism (GITTL, Moscow, 1948).
9. P. Ehrenfest, Proc. Amst. Akad. 36:153 (1933).
10. K. P. Belov, Magnetic Transformations (GIFML, Moscow, 1959).
11. R. C. L. Bosworth, Ferromagnetism [Russian translation] (IL, 1956).
12. V. K. Semenchenko, Kristallografiya, 2:147 (1957).
13. V. K. Semenchenko, Application of Ultra-acoustics to the Study of Matter, Vol. 16 (MOPI, Moscow, 1962), p. 101; Kolloidnyi zhurnal, 24:323 (1962).
14. V. K. Semenchenko and M. M. Martynyuk, Kolloidnyi zhurnal, 24:611 (1962).
15. V. K. Semenchenko, Zh. Fiz. Khim., 36:15 (1962).

FLUCTUATIONS AND THERMODYNAMIC STABILITY
OF SYSTEMS IN THE REGION OF CRITICAL
AND HYPERCRITICAL TRANSFORMATIONS

K. V. Arkhangel'skii

In a number of papers on critical phenomena [1–3], Semenchenko has shown that the fluctuations and thermodynamic stability of various systems are directly interlinked. According to Semenchenko's theory, the thermodynamic stability of a system is characterized by the behavior of its adiabatic and isodynamic coefficients of stability,[*] and the properties of the system by their reciprocals.

We have experimentally checked the conclusions of the theory for binary liquid systems having an upper critical temperature of phase separation.

Before presenting the experimental results showing the effect of fluctuations on the thermodynamic stability of the systems in question in the region of critical and hypercritical transformations, we must decide on two points: How shall we treat the concept of critical temperature, and what is the meaning of the "hypercritical region ?"

We know that the critical temperature is called the "temperature of phase identity." We must not, however, treat it as the temperature at which a visible meniscus appears or vanishes, as do many.[†]

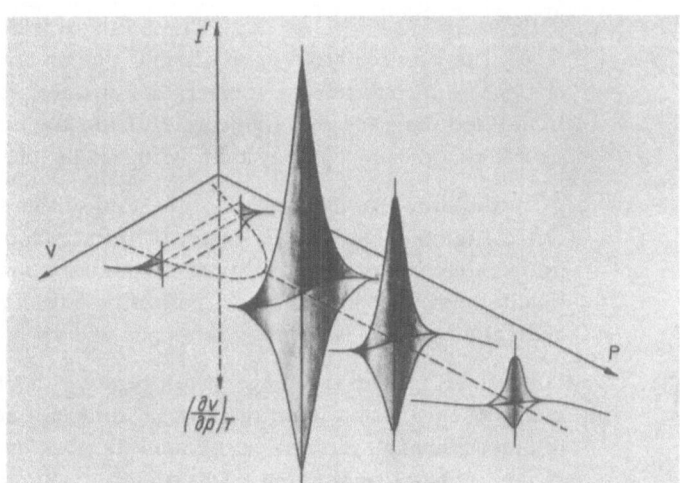

Fig. 1. Geometrical interpretation of the behavior of a simple system (CO_2) in the precritical and hypercritical regions.

---

[*] The second partial derivatives of the thermodynamic potentials with respect to the same coordinate are called adiabatic coefficients of stability if they are taken with all the other generalized thermodynamic coordinates constant, and isodynamic if taken with the generalized thermodynamic forces constant.

[†] In liquid systems.

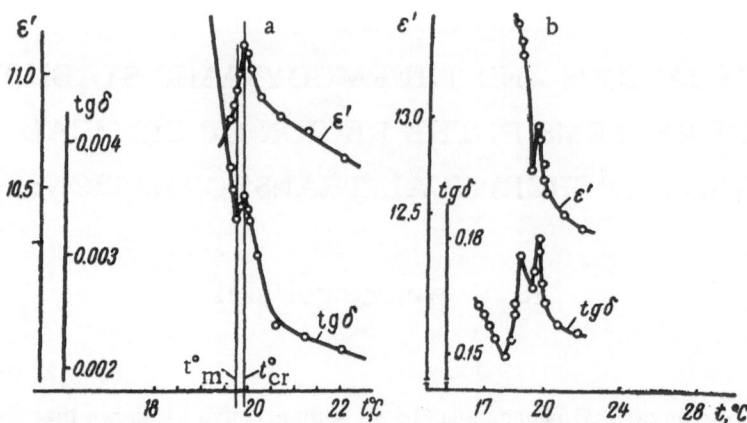

Fig. 2.   Behavior of ε' and tan δ at   the critical point and in the
contiguous regions: a) C = 50. mol. % nitrobenzene in heptane, frequency f = 2 Mc/ sec(t_{cr} = critical temperature, t_M temperature
for the formation of a visible meniscus); b) C = 47.84 mol. %
nitrobenzene in heptane (more than critical), f = 5 kc/ sec.

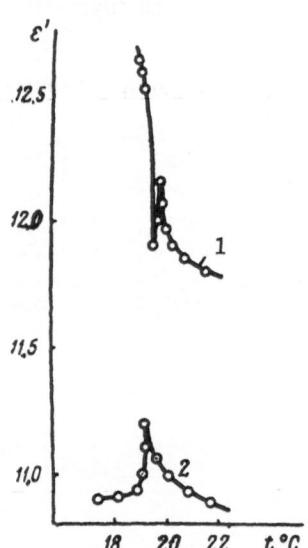

Fig. 3. Effect of concentration
on the temperature-dependence
of the dielectric constant frequency 10kc /sec: 1) C = 47.84
mol.% nitrobenzene in heptane;
2) C = 45.05 mol. % nitrobenzene in heptane.

In support to this we need hardly quote Stoletov, who considered that "the
appearance of a meniscus (or the turbidity preceding it) on cooling takes
place below the critical temperature; at the critical temperature it is impossible by its very definition."* Hence it follows (and this is confirmed
experimentally) that the critical temperature, as the temperature of phase
identity, is not identical with the temperature at which a visible meniscus
appears or disappears.†

The question arises as to whether we can assert, on this basis, that the
critical temperature is the temperature at which there is a transformation of
liquid into gas,‡ and that immediately after the critical temperature the
simple single-component system, being single-phase, becomes gas.

Stoletov thus characterizes the state of this system after reaching the
critical temperature: "The liquid, therefore, on heating above the critical
temperature, loses the capacity to be in mixture with its vapor, and becomes
a quite homogeneous substance, possessing both liquid and gaseous properties,
and called with equal correctness condensed vapor or rarefied liquid. §

In fact, there is a range of temperature in which the substance sub-
sisting in such a state after the critical temperature (the mesophase state
in Semenchenko's terminology) gradually takes on the properties of a gas
as the temperature is raised. This region is also called hypercritical. The
existence of such a region and the transformations taking place in it were
mentioned even in 1876 by Thomas Andrews [7].

Figure 1 shows the geometrical interpretation of the behavior of a
simple system (CO_2) in the precritical and hypercritical (temperature)

---

*A. G. Stoletov. Collected Works (GTTI, Moscow—Leningrad, 1950), p.289.

†This is discussed in more detail in [4].

‡As done in educational literature, see, e.g., [5], pp. 438—39.

§A. G. Stoletov. Collected Works (GTTI, Moscow—Leningrad, 1947), Vol. 3, p. 555.

regions from the data of [8] and [9]. The isotherms of CO_2 in the precritical and hypercritical regions are shown in the P—V plane, and it is also shown (qualitatively) in what way light-scattering takes place and how the isothermal compressibility $(\partial V / \partial P)_T$ of the system varies along these isotherms.

A precritical transformation takes place through the region of instability of the system (indicated in the figure by the dotted curve); the whole sequence of states in this region cannot be realized, one of the phases of the system as it were "pushing" through the region at constant P and T and thereby expending the heat of transformation supplied from outside.

A hypercritical transformation takes place through a region of reduced stability, the stability varying continuously and reaching its minimum value at the inflection in the hypercritical isotherm without disrupting the system, the extrema of the properties of the system in the hypercritical region thus also being finite. It should be noted that at the actual critical point (the inflection point of the critical isotherm), the extrema of the properties can only be infinitely large if they are associated with zero values of the corresponding isodynamic coefficients of stability of the system.

If, however, the extrema of the properties of the system are associated with the variation of its adiabatic coefficients of stability, then they will be finite over the whole critical region, * since the adiabatic coefficients of stability pass through finite extrema at the temperature transformation points for critical and hypercritical transformations, and the greatest of these occurs at the critical point [3]. One of these coefficients of stability is the quantity $1/\varepsilon_a$, where ε_a is the adiabatic dielectric constant of the system under examination. According to the theory in question, which is applicable as well to other than simple single-component systems of the liquid—vapor type, the thermodynamic stability of any system decreases as the critical temperature is approached.

The reduction in the thermodynamic stability of a system is caused by the development of fluctuations within it, which, as Smoluchowski [10] showed, are at a maximum at the critical point. This also leads to the appearance of anomalies in various properties of the system, since these properties are proportional to the fluctuations. The stability coefficients of the system are connected to the fluctuations by a reciprocal law [3], and hence the more the fluctuations are developed in the system, the lower is its stability, and the more clearly do anomalies in the properties appear.

In verifying these conclusions for binary liquid systems, we used the dielectric method of studying the thermodynamical properties of systems in the critical region. As already indicated, one of the stability coefficients is a quantity inversely proportional to the dielectric constant of the system studied. Hence by studying the behavior of the dielectric constant of any system we can obtain information on the behavior of the thermodynamic stability of the same system under given conditions.

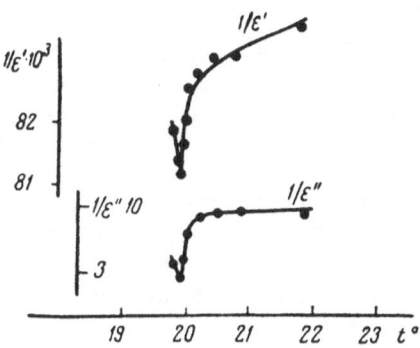

Fig. 4. Temperature-dependence of $1/\varepsilon'$ and $1/\varepsilon''$ for C = 42.95 mol.% nitrobenzene in hexane and frequency f = 5 kc/sec.

A binary liquid system in its critical region undergoes a series of complex transformations. At a temperature considerably higher than critical, the system constitutes an ordinary solution. On approaching the critical temperature, fluctuations can be seen to develop in the system, the more so as the critical temperature is approached. At the critical temperature the fluctuations reach a maximum. After passing the critical temperature the system assumes a colloidal formation, which on lowering of the temperature is transformed into an emulsion with the subsequent appearance of a visible meniscus.

On further lowering of the temperature (for binary liquid systems having an upper critical temperature of phase separation), the drops of emulsion coalesce and there is a further separation of the phases. Under these conditions the difficulty of the experiment is chiefly connected not only with the fact that the principal changes in the system

_____

* The region of critical and hypercritical transformations with respect to temperature are usually known as the critical region.

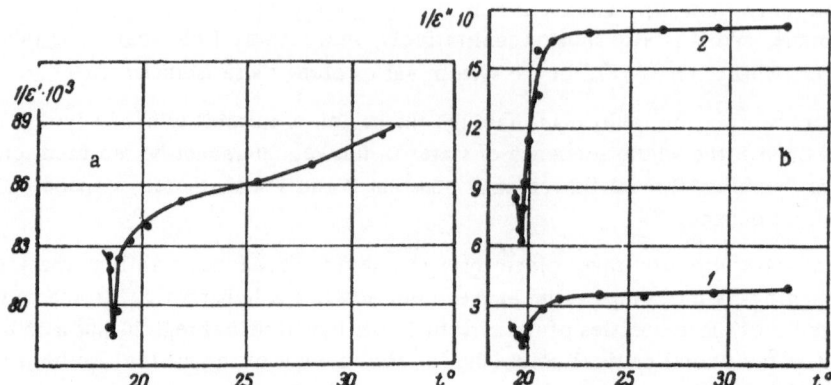

Fig. 5. Temperature-dependence of a) $1/\varepsilon'$ for C = 52.65 mol. % nitrobenzene in octane and frequency f = 5 kc/ sec, and b) $1/\varepsilon''$ for C = 52.65 mol. % nitrobenzene in octane (1: f = 2 kc/ sec; 2: f = 20 kc/ sec).

are occurring in the immediate neighborhood of the critical temperature and in a very narrow temperature range (tenth parts of a degree), but also with the fact that these changes, affecting the behavior of the stability coefficients of the system,[*] themselves depend substantially on gravitation, so that the critical state is realized only in a narrow horizontal layer. In view of this, the heat regulator must have a high "resolving power" with respect to temperature for measurements near the critical temperature, and exact, stable, prolonged thermostating of the sample under examination is required.

It is otherwise easy to pass right through this region without noticing the passage of the stability coefficients through extrema, even with a measuring cell answering the requirements for measuring binary liquid systems in the critical region. We measured the dielectric parameters of the systems selected in a special condenser with horizontally placed plane electrodes separated by 1.2 mm, using accurate, stable thermostats with a precision exceeding 0.01°C.

The condenser has no other volume than that between one of the electrodes and the insulator associated with the other electrode.

In order to elucidate how the very fact of phase separation and emulsion formation affected the behavior of ε' and tan δ   in the critical region, we studied the behavior of nitrobenzene—heptane mixtures with concentrations greater and less than critical.

After phase separation of the system, in both cases the liquid dielectric proved inhomogenous.

For high-frequency measurements, the ε' of this system, after the appearance of the visible meniscus, continued to decrease in both cases, but for low-frequency (sonic) measurements the ε' of a system with more than critical nitrobenzene concentration increased, while that of a system with less than critical nitrobenzene concentration fell, though in neither case was the system homogeneous.

It is also known that inhomogeneity does not introduce loss by itself. "There are no extra 'Wagner' losses caused by the inhomogeneity of the dielectric. The structure of the dielectric governs only part of the participation of various components in the total energy scattering."[†]   Some results of the measurements in question are given below. Figures 2 and 3 give the temperature-dependence of the dielectric parameters of the nitrobenzene—heptane system with concentrations more and less than critical and measurements in the radio- and audio-frequency ranges.

Figures 4 and 5 show how the thermodynamic stability of nitrobenzene—hexane and nitrobenzene—octane falls as the critical temperature is approached as well as during the transformation itself.

---

[*] And especially the dielectric parameters of the system, characterized in variable fields by the quantities $1/\varepsilon'$ and $1/\varepsilon''$.

[†] Physics of Dielectrics [Russian translation], Editor, A. F. Walter (GTTI, 1932), p. 236.

The results of our investigation lead to the following conclusions:

1. At concentrations close to critical, the dielectric parameters of the system pass through maxima at the critical temperature. The positions of the maxima coincide with respect to temperature and do not depend on frequency.

2. The critical temperature (temperature of phase identity) at which the maxima of ε' and $\tan\delta$ are found does not coincide with the temperature at which a visible meniscus is formed.

3. Qualitatively, the behavior of the dielectric parameters of the system at different concentrations differs only after formation of a visible meniscus.

4. The relative magnitude of the ε' and $\tan\delta$ extrema increases with falling frequency, other conditions being equal.

5. The thermodynamic stability of the system may be studied by the dielectric method.

6. The fact that the dielectric parameters of the system rise and pass through a maximum at the critical temperature on approaching it from the side of the single-phase state, while the stability coefficients decrease, reaching a minimum at the critical temperature (as clearly seen in Figs. 4 and 5), indicates that the conclusions of the generalized theory of phase equilibria and transformations developed by Semenchenko are in full agreement with the experimental data.

Literature Cited

1. V. K. Semenchenko, Zh. Fiz. Khim., 21: 1461 (1947).
2. V. K. Semenchenko, Kristallografiya, 2:145 (1957).
3. V. K. Semenchenko, Zh. Fiz. Khim., 35: 2448 (1961).
4. K. A. Arkhangel'skii and V. K. Semenchenko, Zh. Fiz. Khim., 36: 2564 (1962).
5. K. A. Putilov, Physics Course, Ch. 1 (Moscow, 1959).
6. A. G. Stoletov, Collected Works, Vol. 3, (GTTI, Moscow—Leningrad, 1947).
7. T. Andrews, Continuity of the Gaseous and Liquid States of Matter [Russian translation from English] (GTTI, 1933).
8. A. Michels, B. Blaisse, and C. Michels, Proc. Roy. Soc., A160:358 (1937).
9. V. P. Skripov and Yu. D. Kolpakov, Collection: "Critical Phenomena and Fluctuations in Solutions" [in Russian] Izd. Akad. Nauk SSSR, 1960, p. 126.
10. M. Smoluchowski, Ann. Phys., 25:205 (1908).
11. K. V. Arkhangel'skii and V. K. Semenchenko, Zh. Fiz. Khim., 36:2501 (1962).

PHASE TRANSFORMATIONS OF THE SECOND KIND*

N. N. Sirota

Phase transformations of the second kind [1] are extremely widespread in nature. To these belong the phenomena of ordering, discovered by Kurnakov, Zasedatelev, and Zhemchuzhnyi for Cu—Au alloys [2], and Urazov in Mg—Cd alloys [3]. In later years, ordering phenomena were observed in many other metal alloys [4] and nonmetallic systems. To phase transformations of the second kind also belong ferromagnetic and ferro-electric transformations in the neighborhood of the Curie temperature, transformations in liquid helium, and transformations from the nonsuperconducting to the superconducting state.

The problem of phase transformations of the second kind is one of a number of important problems on the physics of condensed media. Up to the present there has been no strict theory of phase transformations. Fruitful results in this sphere of reach were obtained by Ehrenfest [1], Keesom [5], and Landau and Lifshits [6]. The theory of Landau and Lifshits, developed by Ginzburg et al., has consituted an important contribution, although it contained a number of imprecise concepts. Extensive and important studies in regard to critical phenomena and phase transformations were carried out by Semenchenko [7]. Certainly Semenchenko's work on the analysis of the stability determinant and the fluctuation phenomenon should have a substantial effect on the further development of the subject.

Papers by Leontovich [8] and Ioffrei [9] contained valuable results and gave a general criticism of the incorrect concepts of existing theories. As a whole, however, the theory of phase transformations of the second kind is in an early stage of development.

In contrast to phase transformations of the first kind, for which, in single-component substances at constant pressure, nonvariant equilibrium exists, phase transformations of the second kind extend over a certain temperature range.

Phase transformations of the first kind in single-component systems with one variable external parameter are characterized by an intersection point of the thermodynamic potential curves of the phases respectively existing at lower and higher values of the variable external parameter (temperature T, pressure P, magnetic field strength H, electric field E) than the value corresponding to the intersection (T_K, P_K, H_K, E_K) (Fig. 1). Moreover, in phase transformations of the first kind it is assumed that the thermodynamic potential curves of the phases can be extended into metastable regions beyond the intersection point, so that phases in a metastable state (for example, superheating and supercooling) really can exist.

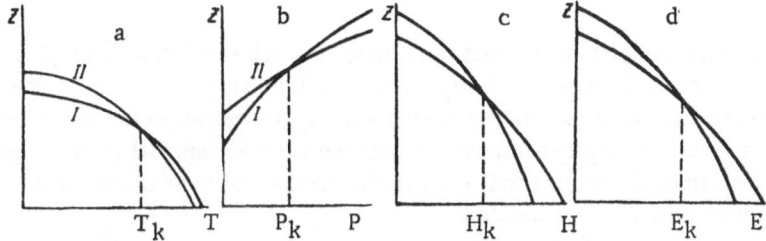

Fig. 1. Variation of the thermodynamic potential of phases I and II in the region of a phase transformation of the first kind a) with temperature, b) with pressure, c) with magnetic and d) electric field strength.

*The main principles of the paper were presented to the Symposium on Ferromagnetism in Krasnoyarsk in 1962 and to the Third Conference on Ferrites in Minsk.

The thermodynamical condition of equilibrium at the transformation temperature T_K is the equality of the chemical potentials of the respective phases μ_I and μ_{II}, and, since we are concerned with a single-component system, correspondingly the equality of the thermodynamic potentials:

$$Z_1 = Z_2,$$
$$dZ_1 = dZ_2.$$

The thermodynamic potential is a function of temperature and the generalized forces $Z = Z(T, X_i)$. Hence the equality $Z_1 = Z_2$ is equivalent to the condition

$$\left(\frac{\partial Z_1}{\partial T}\right)_{X_i} dT + \sum \left(\frac{\partial Z_1}{\partial X_i}\right)_T dX_i = \left(\frac{\partial Z_2}{\partial T}\right)_{X_i} dT + \sum \left(\frac{\partial Z_2}{\partial X_i}\right)_T dX_i.$$

If only one of the generalized forces P, H, or E varies, then

$$- S_1 dT + \varkappa_{i, 1} dX_{i, 1} = - S_2 dT - \varkappa_{i, 2} dX_{i, 2},$$

whence follows the generalized Clausius–Clapeyron equation

$$\frac{dX_i}{dT} = \frac{S_2 - S_1}{\varkappa_2 - \varkappa_1},$$

which is valid for phase transformations of the first kind. For phase equilibrium at $T = T_K$,

$$S_2 - S_1 = \frac{\Delta H}{T_\kappa}.$$

Then

$$\frac{dX_i}{dT} = \frac{\Delta H}{T_\kappa (\varkappa_{2, i} - \varkappa_{1, i})}.$$

At the phase transformation point of the first kind there thus arises a latent heat of transformation ΔH. On transforming from phase I to phase II there is a finite change of volume, entropy $\Delta S_{2,1}$, or corresponding generalized coordinate (magnetic moment, electrical polarization). According to the Gibbs phase rule, a phase transformation of the first kind in a single-component substance for one external variable parameter is a nonvariant transformation, i.e., in equilibrium conditions, in the absence of heat inflow or outflow, equilibrium corresponds to a zero number of degrees of freedom.[*]

In contrast to phase transformations of the first kind, phase transformations of the second kind are characterized by a gradual change from one state (initial phase) to a second state (nascent phase) through a continuous array of an infinitely large number of intermediate states in a transformation range corresponding to a definite range of variation in the external equilibrium parameter.

---

[*] A nonvariant transformation of the first kind in a single-component system taking place at constant external pressure extends into a range of variation of the external equilibrium parameter if an extra condition, such as constancy of volume, is applied.

Thus, in this range of variation of the external parameter (in the transformation range), to every given value of the varying external equilibrium parameter there corresponds unambiguously a definite value of the resultant thermodynamic potential of the system. Hence in the transformation range, for any value of temperature, pressure, or other external equilibrium parameter, there is no latent heat of transformation or finite change of entropy, volume, or any generalized coordinate, i.e.,

$$\left(\frac{\partial \Delta Z}{\partial T}\right)_{X_i} = 0,$$

$$\left(\frac{\partial \Delta Z}{\partial X_i}\right)_T = 0,$$

whence, as we know, Ehrenfest's equation [1] follows immediately, since

$$\frac{\partial \Delta Z}{\partial T} = \frac{\partial^2 \Delta Z}{\partial T} dT + \frac{\partial^2 \Delta Z}{\partial T \partial X_i} dX_i = 0,$$

$$\frac{\partial \Delta Z}{\partial X_i} = \frac{\partial^2 \Delta Z}{\partial X_i \partial T} dT + \frac{\partial^2 \Delta Z}{\partial X_i^2} dX_i = 0$$

and hence

$$\begin{vmatrix} \dfrac{\partial^2 \Delta Z}{\partial T^2} & \dfrac{\partial^2 \Delta Z}{\partial T \partial X_i} \\[2ex] \dfrac{\partial^2 \Delta Z}{\partial X_i \partial T} & \dfrac{\partial^2 \Delta Z}{\partial X_i^2} \end{vmatrix} = 0$$

or

$$\Delta C_\Gamma = \frac{T \left(\dfrac{\partial^2 \Delta Z}{\partial X_i \partial T}\right)^2}{\dfrac{\partial^2 \Delta Z}{\partial X_i^2}}.$$

If we dismiss the trivial case, then the variation of the resulting thermodynamic potential of a single-component system in the transformation range is a single-valued function of the external equilibrium parameter and the internal varying parameter [degree of order, magnetic moment, polarization $Z = Z(T, X_i, \eta)$]. For phase transformations of the second kind, Gibbs' phase rule cannot be directly applied. Allowing for the variation of the internal parameter in the single-component system, the number of degrees of freedom in this case will be greater than zero (for example, for variation of T and P = const.) [10].

Phase transformations of the second kind, such as transformations from one state (one phase) to another state (second phase) through an infinite series of intermediate equilibrium states, may be regarded as transformations from one phase to the other through an infinite series of intermediate equilibrium phases (infinite number of polymorphic modifications), each which corresponds to a definite value of the internal parameter η.

The variation of the thermodynamic potential of each of the phases for constant internal parameter (η = const.) may be represented as a general functional dependence on temperature, pressure, or some other generalized force:

$$Z = Z_0 - k f(T), \quad Z = Z_0 + b \, \varphi(P).$$

First of all, let us introduce, instead of temperature, the function $\xi = f(T)$ or, instead of pressure, the function $\psi = \varphi(P)$. The variation of the thermodynamic potential of each of the phases as a function of ξ or φ will be linear:

$$Z = Z_0 - k\xi,$$
$$Z = Z_0 + c\psi.$$

From this it is evident that k and, correspondingly, c are the angular coefficients, equal to the tangents of the angle of inclination of the straight lines:

$$-k = \operatorname{tg}\alpha = \frac{\partial Z}{\partial \xi}, \quad c = \operatorname{tg}\beta = \frac{\partial Z}{\partial \psi}.$$

Figure 2 shows the case of a transformation from phase I to phase II for variation of ξ through five intermediate phases (2, 3, 4, 5, 6); this may be considered as the case of a substance with seven polymorphic modifications. Hence phase I exists as equilibrium phase in the range of ξ values from 0 to ξ_1, phase 1 in the range ξ_1 to ξ_2 (T_1 to T_2), phase 2 from ξ_2 to ξ_3 (T_2 to T_3) . . ., and phase II for $\xi > \xi_{II}$.

The resulting variation of the equilibrium thermodynamic potential of the system in the range ξ_I to ξ_{II} will be described by a broken line formed from the five parts mentioned.

Figure 3 shows the case in which a single-component substance contains not only phases I and II but also five metastable phases (2, 3, 4, 5, 6) which are not stable for any values. Here the resulting variation in the equilibrium thermodynamic potential of the system will be described by two straight lines intersecting at the point $\xi_{I, II}$ ($T_{I, II}$).

As the number of intermediate equilibrium phases with corresponding value of internal parameter η (see Fig. 2) rises to infinity for a continuous transformation, to each phase on the resulting thermodynamic potential curve of the system in the $\xi_I - \xi_{II}$ transformation range will correspond a certain segment constricted to a single point. The resultant thermodynamic potential curve will thus form the envelope of a family of thermodynamic potential curves of an infinitely large number of intermediate equilibrium phases (for example, an infinitely large number of polymorphic modifications), densely filling the range from ξ_I to ξ_{II}. To each point on this envelope (curve of thermodynamic potential of the system as a function of ξ) will correspond a certain state ("phase") with a certain internal parameter value η.

On the other hand, each state representing a given value of internal parameter η has its corresponding thermodynamic potential curve, extending from zero to specific values of the generalized force over the whole metastable range, and passing through the stable point lying on the envelope.

A necessary condition in order for each point of the envelope (resultant thermodynamic potential curve) in the transformation interval to correspond to an equilibrium intermediate state, is that the convexity of the resultant curve should go upward from the axis (see Fig.2). On the other hand, if there is a continuous set of intermediate metastable phases and hence only a stable transformation from phase I to phase II, passing by the intermediate state, the resultant curve is concave, or rather the convexity is downward, toward the axis of abscissas. Thus the first case, corresponding to phase transformations of the second kind, differs from the second case with respect to the envelope derivative $\partial^2 Z / \partial \xi^2$.

The change in the direction of convexity of the envelope also depends on the connection between the thermodynamic potential at $T = 0$, i.e., on the zero enthalpy close to the automization energy $Z_0 = \Delta H_0$, with the angular coefficient $\partial Z / \partial \xi$. In the case when the thermodynamic potential envelope is convex–upward (phase transformations of the second kind), the variation of the angular coefficient

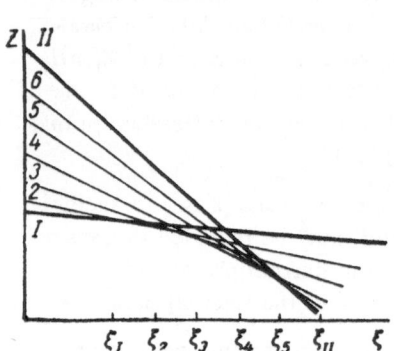

Fig. 2. Variation of thermodynamic potential with ξ for phase I and II and five intermediate equilibrium polymorphic forms.

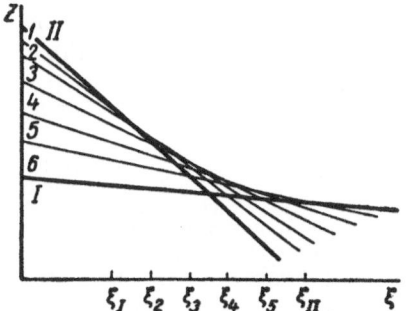

Fig. 3. Variation of thermodynamic potential with ξ for phase I and II and five intermediate metastable forms.

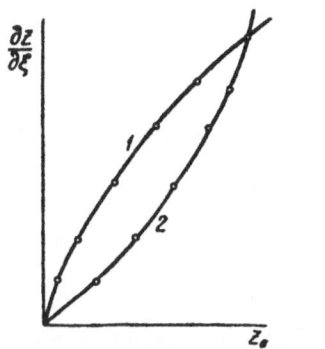

Fig. 4. Variation of $\partial Z/\partial \xi$ with Z_0 for 1) convex—upward and 2) convex—downward resultant thermodynamic potential curves in the transformation range, also depending on ξ.

with Z_0 is also characterized by upward convexity (Fig. 4, curve 1). If the thermodynamic potential envelope is convex—downward, then the curve of angular coefficient against Z_0 is also convex—downward (Fig. 4, curve 2).

Thus the angular coefficient k is a function of the zero thermodynamic potential $\varphi(Z_0)$, and the sign of its second derivative with respect to Z_0 characterizes the sense of the convexity of the resultant (envelope) thermodynamic potential curve in the transformation interval.

Moreover, the value of Z_0 for each intermediate phase is determined entirely by the variation of the internal parameter η. Since, for each intermediate phase, for a given value of η, the thermodynamic potential is a linear function of $\xi(T)$ or $\psi(X_i)$, and the angular coefficients k are functions of Z_0, if we confine consideration to the relation between Z and ξ, we shall have the system of equations

$$Z - Z_0 + k\xi = 0,$$
$$k - a\psi(Z_0) = 0$$

or

$$\varphi = Z - Z_0 + a\psi(Z_0)\xi = 0.$$

Then the equation of the envelope will be found from the condition

$$\frac{\partial \varphi}{\partial Z_0} = -1 + a\frac{\partial \psi(Z_0)}{\partial Z_0}\xi = 0.$$

The functional relation $k = a\psi(Z_0)$ may be taken on the basis of physical considerations. In the simplest case it may be approximated by a power function of the type $k = aZ_0^m$. In this case $Z = Z_0 - aZ_0^m\xi$, and the equation of the envelope in the transformation range from ξ_I to ξ_{II} will have the form

$$Z = \left(\frac{1}{am}\right)^{\frac{1}{m-1}}\left[1 - \frac{1}{m}\right]\xi^{-\frac{1}{m-1}}.$$

(In Fig. 2 the envelope is given for the case in which $m = 2/3$). In this case the derivatives of Z with respect to ξ in the transformation range will equal

$$\frac{\partial Z}{\partial \xi} = -\frac{1}{m-1}\left(\frac{1}{am}\right)^{\frac{1}{m-1}}\left[1 - \frac{1}{m}\right]\xi^{-\frac{m}{m-1}},$$

$$\frac{\partial^2 Z}{\partial \xi^2} = \frac{m}{(m-1)^2}\left(\frac{1}{am}\right)\left[1 - \frac{1}{m}\right]\xi^{\frac{1-2m}{m-1}}.$$

The variation of thermodynamic potential and the first derivative $\partial Z/\partial \xi$ is shown in Fig. 5 (curves 1 and 2). Here we have also given the curve for the product of the second derivative and the coordinate ($\partial^2 Z/\partial \xi^2)\xi$ (Fig. 5, curve 3).

In the Debye approximation, at low temperatures ξ is proportional to the fourth power of temperature ($\xi \sim T^4$), and at high temperatures to the first power ($\xi \sim T$).

At sufficiently low temperatures, in the Debye approximation, the lattice free energy will be proportional to the fourth power of temperature divided by the cube of the characteristic temperature. Since the specific heat

$$C_v \sim C' \left(\frac{T}{\Theta}\right)^3,$$

then

$$\Delta Z \sim F = -T \int \frac{dT}{T^2} \int C' \left(\frac{T}{\Theta}\right)^3 dT = CT \left(\frac{T}{\Theta}\right)^3 =$$

$$= 0.0386\, T \left(\frac{T}{\Theta}\right)^3.$$

On the other hand, it is known that the characteristic temperature is approximately proportional to the square root of the bulk modulus over the atomic weight:

$$\Theta = \frac{h}{k} \sqrt{\frac{\varkappa}{A}}.$$

The bulk modulus, in turn, according to Gruneisen, may be regarded as proportional to the heat of atomization at zero temperature, i.e., the thermodynamic potential at absolute zero:

$$\varkappa = \frac{mn Z_0}{V},$$

where m and n are the powers in the Mee—Gruneisen Law. Thus the relation for the thermodynamic potential at absolute zero may be represented as

$$Z_0 = \gamma \frac{A \Theta^2}{mn}$$

or, assuming that the product mn remains approximately constant in the transformation interval as the internal parameter varies, in the simplest case

$$Z_0 = A \alpha \Theta^2.$$

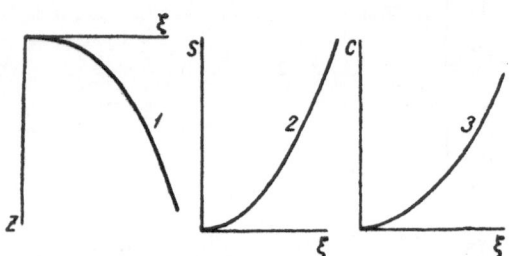

Fig. 5. Thermodynamic potential(envelope) as a function of ξ in the transformation range, first derivative $-\partial Z/\partial \xi$, and $(\partial^2 Z/\partial \xi^2)\xi$.

The quantity α is defined as the ratio of the atomization energy or Z_0 to the product of the square of the characteristic temperature and the atomic weight.

For simple substances with the same valence and crystal lattice type we may say that α is approximately constant [11, 12]. For example, for silicon $\alpha_{Si} = 0.92 \cdot 10^{-2}$, for germanium $\alpha_{Ge} = 0.91 \cdot 10^{-2}$. However, even for iron $\alpha = 0.93 \cdot 10^{-2}$, and for cobalt α_{Co} $1.15 \cdot 10^{-2}$. Let us take $\alpha = 1 \cdot 10^{-2}$.

We may suppose that α will remain practically unchanged for different values of internal parameter in one and the same substance.

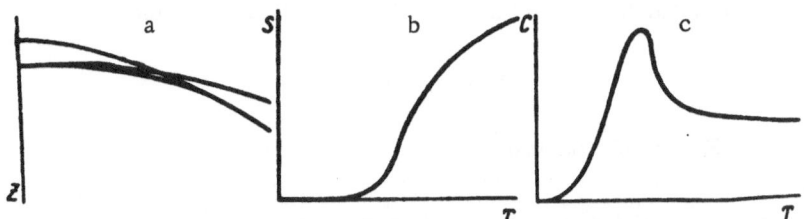

Fig. 6. Variation of a) thermodynamic potential, b) entropy, and
c) specific heat in the transformation region for the case in which
$Z(T)$ at $\eta = $ const. is described by the Planck–Einstein relation [for
$\Delta Z_0 = - b \ln (\Theta/\Theta_1)$].

In this case, the course of the thermodynamic potential curve for constant internal parameter can be
represented in the low-temperature region by the following relation:

$$Z = Z_0 - CT \left(\frac{T}{\Theta} \right)^3 = A \alpha \Theta^2 - CT \left(\frac{T}{\Theta} \right)^3 .$$

Eliminating the parameter (characteristic temperature) from the condition

$$\frac{\partial}{\partial \Theta} \left[Z - A \alpha \Theta^2 + C \frac{T^4}{\Theta^3} \right] = 0,$$

we find the equation of the envelope (thermodynamic potential curve) in the transformation interval:

$$Z = - 2 C^{0.4} \alpha^{0.6} A^{0.6} T^{1.6}.$$

Then in the transformation interval the entropy variation will be

$$- \frac{\partial Z}{\partial T} \simeq 2 C^{0.4} \alpha^{0.6} A^{0.6} \frac{8}{5} T^{3/5}$$

and the specific heat variation will be

$$- \frac{\partial^2 Z}{\partial T^2} T \simeq 2 C^{0.4} \alpha^{0.6} A^{0.6} \frac{24}{25} T^{3/5}.$$

As an example of applying the method here developed for analyzing the thermodynamic potential curves in the
neighborhood of a phase transformation of the second kind, let us also consider the case in which the thermodynamic potential curves of the intermediate states, for a constant value of an internal parameter (denoted for
example, by the characteristic temperature), may accurately be fairly represented by Planck–Einstein functions.

In this case, as we know,

$$U = N \frac{h\nu}{e^{\frac{h\nu}{kT}} - 1},$$

where

$$\frac{h\nu}{k} = \Theta.$$

Assuming, as in the previous case, the validity of the approximate relation

$$\Delta Z_0 = - b \Delta(\Theta)^2$$

and eliminating parameter Θ from the condition

$$\frac{\partial}{\partial \Theta} \left[\Delta Z + b \Delta(\Theta)^2 - 3RT \ln \left(1 - e^{-\frac{T}{\Theta}} \right) \right] = 0,$$

we find the equation of the envelope, i.e., the resultant variation of the thermodynamic potential of the system in the region of transformation:

$$\Delta Z = A + 3RT \ln \left(1 - e^{-\frac{3R}{2bT}} \right).$$

The entropy and specific heat variation in the transformation range will, moreover, be equal to

$$S = 3R \left[\frac{\frac{3R}{2bT}}{e^{3R/2bT} - 1} - \ln \left(1 - e^{-\frac{3R}{2bT}} \right) \right],$$

$$C = 3R \frac{\left(\frac{3R}{2bT} \right)^2 e^{3R/2bT}}{(e^{3R/2bT} - 1)^2}.$$

In the limiting case, the value of ΔZ_0 may be connected with $\Delta \Theta$ by the relation $\Delta Z_0 = - b \ln (\Theta/\Theta_1)$.

Figure 6 shows the variation of S, C, and Z in the range of a phase transformation of the second kind for the case where the temperature-dependence (of the specific heat of each of the infinite number of intermediate states, for constant value of the internal parameter) is described by the Planck—Einstein relation and $\Delta Z_0 = -b \ln (\Theta/\Theta_1)$. Analogous relations may be obtained on the approximate temperature/specific heat theory of Debye et al.

Phase transformations of the second kind, from the original state I to the final state II, are described by a thermodynamic potential curve which in the transformation region cannot be represented as any combination of the thermodynamic potential curves of the original and final phases (intersection with contact, and so forth [13]). Each point of the equilibrium resultant curve in the region of the phase transformation (i.e., each point of the envelope considered above) corresponds to an intermediate equilibrium thermodynamic state of the system. Hence from this point of view a phase transformation of the second kind is not characterized by points (of temperature, pressure, magnetic field strength, etc.) but by an interval of transformation as one of the external equilibrium factors varies. Thus, in general, phase transformations of the second kind must have points corresponding to the beginning and end of the transformation (beginning and end of the thermodynamic potential envelope in the transformation interval). The end point (T_c) of the transformation in a single-component system, for variation of (for example) the temperature, in the majority of cases corresponds to the temperature of a phase transformation of the second kind, i.e., the Curie temperature of ferroelectrics and ferromagnetics, or the Kurnakov temperature in ordered alloys. In many cases the starting temperature of the transformation (T_0) may coincide with absolute zero. In principle, however (considering the question of from the viewpoint of geometric thermodynamics), this is not obligatory, and is entirely determined by the functional dependence of the entropy and magnitude Z_0 of the zero free energy on the internal parameter.

As seen from the above examples, the equation of the envelope is in many cases close to the equation of a second-order curve. In a number of cases the envelope may be approximated by the equation of an ellipse. For example, in the transformation range the envelope may be approximated by an equation of type

$$Z = b \sqrt{1 - \left(\frac{T}{a}\right)^n}.$$

Furthermore, the variation of entropy S and specific heat C in the transformation interval is described by the equations

$$S = \frac{nb}{2} \frac{T^{n-1}}{a^n} \left[1 - \left(\frac{T}{a}\right)^n\right]^{-1/2},$$

$$C = \frac{nb(n-1)}{2} \frac{T^{n-1}}{a^n} \left[1 - \left(\frac{T}{a}\right)^n\right]^{-1/2} \times$$

$$\times \left\{1 + \frac{n}{2(n-1)} \left[\left(\frac{T}{a}\right)^n \left[1 - \left(\frac{T}{a}\right)^n\right]^2\right\}$$

or, for n = 2,

$$S = \frac{bT}{a} [a^2 - T^2]^{-1/2},$$

$$C = abT (a^2 - T^2)^{-3/2}.$$

For the direction of convexity, finally, it is useful also to consider the magnitude and sign of the third and fourth derivatives.

The temperatures at the beginning and at the end of the transformation may be determined from the initial and final values of the internal parameter η_I and η_{II} on the basis of the law of variation of $Z_0(\eta)$ as a function of η, and also from the values of entropy for the initial and final states.

In the example considered, when $S = \varepsilon$ is placed at the end of the transformation, the transformation temperature (end of transformation) will equal $T_c = (\varepsilon a/b)$.

Moreover the value of the jump in specific heat equals

$$\Delta C_p = \frac{\varepsilon}{a} \left[1 - \frac{\varepsilon^2}{b^2}\right]^{-3/2} = \frac{bT_c}{a^2} \left[1 - \left(\frac{T_c}{a}\right)^2\right]^{-2/3}.$$

Phase transformations of the second kind, described by the thermodynamic potential envelope in the transformation interval, may vary considerably as regards the way in which the entropy and specific heat vary in the transformation interval, if small changes are made in the shape of the envelope while still satisfying the conditions for the transformation to be thermodynamically possible. Figure 7 shows some examples of the way in which the thermodynamic potential, entropy, and specific heat may vary in the transformation interval; this does not exhaust all the possible different forms. Any limitation to the possible types of transformation of the second kind must be based on an analysis of the functional dependence of the internal parameter, the initial zero value of the thermodynamic potential, the characteristic temperature, and the entropy, which determine the course of the thermodynamic potential curve for a given constant value of the internal parameter.

For many of the various types of phase transformation of the second kind, the variation of entropy and specific heat with temperature in the transformation interval may be conveniently approximated not so much by functions of parabolic form as by functions representing S-shaped curves [14], for example of the form

$$S = A (1 - e^{-kT^m}),$$

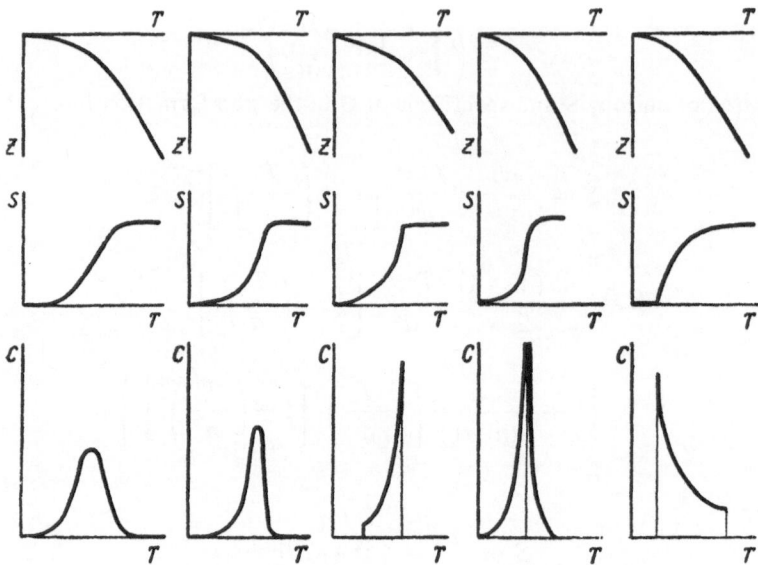

Fig. 7. Examples of variation of Z, S, and C for various types of
transformation in the transformation range.

$$S = aX^n - bX^{qn}.$$

or by logarithmic functions.

Experimentally, it is not always possible to track the variations in thermodynamic potential of a system for constant values of the internal parameter in the metastable region. It is, however, always possible experimentally to determine the constants coming into the equation of the thermodynamic potential envelope and the actual form of the thermodynamic potential curve in the transformation interval, and also partly to determine the variation of Z_0 as a function of the internal parameter, at least at low temperatures. The determination of such data certainly makes it possible to check the theory and to derive the constants needed for deducing a great number of consequences of scientific and practical value. The character of the thermodynamic potential curve (envelope) in the transformation interval indicates that fluctuation processes and transformations through quasiheterogeneous states may develop. The concepts regarding the thermodynamics of phase transformations of the second kind here outlined may be useful for further development of the theory as well as experimental investigations of this important and interesting problem.

Literature Cited

1. P. Ehrenfest, Communications Leiden Suppl., p. 756 (1933).
2. N. S. Kurnakov, M. Zasedatelev, and S. A. Zhemchuzhnyi, Izv. Peterburksk: Politekhn. Inst., p. 485 (1914).
3. G. G. Urazov, Zh. Ross. Fiz. Khim. Obshch., Ser. Khim., 42—I 4:728 (1910).
4. W. Hume-Rothery and G. V. Raynor, The Structure of Metals and Alloys (London, 1962).
5. W. H. Keeson, Communications Leiden Suppl. (1933).
6. L. D. Landau and E. M. Lifshits, Statistical Physics (GITTL, 1957); Zh. Eksperim. i Teor. Fiz. (1937).
7. V. K. Semenchenko, Present Collection, p. 3.
8. M. A. Leontovich, Introduction to Thermodynamics (GITTL, Moscow, 1959); V. L. Ginzburh, Zh. Eksperim. i Teor.Fiz. (1945); 17 (1947).
9. J. Joffrei, Ann. Phys., 3 (5), (1948).
10. V. A. Sokolov, Zh. Eksperim. i Teor. Fiz. 3 (6) (1948).

11. N. N. Sirota, Izv. Sket.Fiz. Khim., Anal.Otd. Khim. Nauk,Akad. Nauk SSSR XXI: 90 (1951).

12. A. A. Vorob'ev, Physical Properties of Ionic Crystal Dielectrics (Tomsk, 1960).

13. Epstein, Course of Thermodynamics [Russian translation] (Ogiz, 1948).

14. N. N. Sirota, Tr. Nauchn. Tekhn. Obshch. Chernoi Met., Vol. VI (1955).

EFFECT OF CERTAIN ADDITIONS ON THE PHYSICO-CHEMICAL PROPERTIES OF BARIUM TITANATE SINGLE CRYSTALS AT PHASE TRANSFORMATION POINTS AND IN THE FERROELECTRIC REGION

M. L. Sholokhovich, A. L. Khodakov,[*] T. N. Lezgintseva, and L. M. Berberova

Systematic investigations on the production and study of ferroelectric solid solution single crystals have resulted in the production of crystals with assigned and, in a number of cases, practically important values of phase transformation points. In particular, there is great practical interest in crystals with high Curie points together with high nonlinearity and rectangular hysteresis loops. From this point of view, crystals of the $BaTiO_3$—$BaHfO_3$ system are singled out among ferroelectric solid solution single crystals [1, 2]. In studying these single crystals, our attention was drawn to the fact that extremely small quantities of $BaHfO_3$, introduced into the crystals in the form of impurities of the order of 0.01 to 0.05%, and hardly affecting the phase transformation temperature, raised all the ferroelectric properties of the crystals extremely sharply, and in particular their nonlinearity. In view of this, it was of special interest to study the effects of very small additions of various substances on the ferroelectric properties of crystals and to elucidate the mechanism of these effects. In this respect there have already been a number of papers [3—5] which have revealed certain additives considerably reducing the ferroelectric properties of crystals.

A marked example of this arises on introducing Fe_2O_3 into barium titanate crystals [3]. A number of papers by Zheludev and Puzyrev have also been devoted to a study of the properties of barium titanate crystals as a function of impurities introduced.

The present paper contains results of studies of the effect of small impurities of the dioxides of elements in the fourth group of the periodic table on the form [6] and the electrical properties of barium titanate single crystals. In the systems $BaTiO_3$—$BaZrO_3$, $BaTiO_3$—$BaSnO_3$, and $BaTiO_3$—$BaHfO_3$, studied both in the form of ferroelectric ceramics and single crystals, the formation of solid solutions has been established [7—12]. Tetragonal solid solutions have also been shown to form in ceramic $BaTiO_3$—ZrO_2 and $BaTiO_3$—HfO_2 [15]. From this we might expect that, on introducing extremely small quantities of ZrO_2, HfO_2, and SnO_2 into $BaTiO_3$ single crystals, the T^{+4} ions in the barium titanate crystal lattice might be replaced by Zr^{+4}, Hf^{+4}, Sn^{+4}, ions the ionic radii of these elements being very similar. As regards additions of SiO_2 and GeO_2, i.e., oxides of elements with considerably smaller ionic radii, the possibility of introducing Si^{+4} and Ge^{+4} ions into the crystal lattice of barium titanate even up to the present time remains uncertain. The same applies to Th^{+4}, which has a considerably larger ionic radius than does Ti^{+4}.

Crystals were grown from a solution of barium titanate with added impurity in a potassium fluoride melt. The original barium titanate was the same in all the experiments, being prepared from $BaCO_3$ and TiO_2 by a solid state reaction. The barium carbonate was prepared by precipitating with CO_2 gas from a solution of chemically pure $Ba(OH)_2$. The "analytically pure" titanium dioxide was further purified by prolonged boiling in a 10% chemically pure acid, with subsequent washing out of the chlorine ion.

---

[*] Deceased.

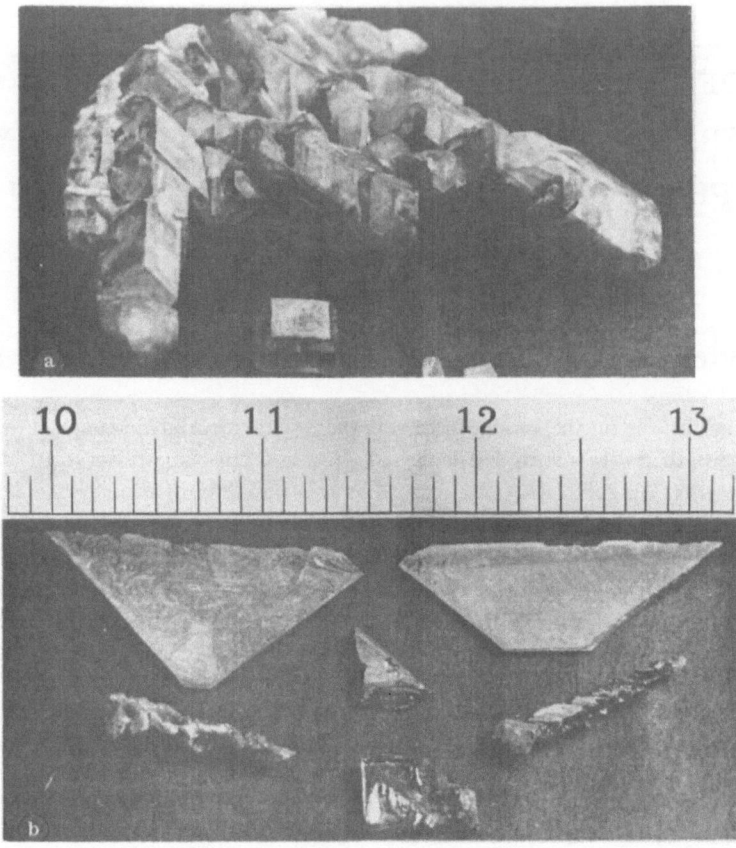

Fig. 1. Form of barium titanate crystals grown from melt containing a) SiO_2 and b) ZrO_2.

The impurities were in all cases introduced into the original melt in the identical proportion of 1 mol. %.
Experiments on growing crystals with different additives were made under completely identical conditons, both
as regards the concentrations of the original melt and as regards temperature conditions. The experimental
conditions are described in [16].

Spectral analysis of the single crystals obtained showed that different additives penetrated into the crystals
in very small and unequal quantities of the order of 0.01 to 0.1 wt. %. In crystals grown from original mixtures
containing SiO_2, the silicon content was rather greater (up to 1 wt. %). The impurities introduced under these
growth conditions exerted a considerable influence on both the crystallization process and the shape and total
yield of the crystals (Fig. 1).

Upon the introduction of ZrO_2 impurity into the $BaTiO_3$, some of the crystals emerged in the form of
very small, poorly faced crystals of cubic shape, with a small number in the form of nodules. Together with
these were a considerable number of crystals in the form of flat plates in the shape of trapezia, growing together along the most developed edge of the facets into twins at an angle of less than 45°. Most of the plates
had edges up to 0.6 cm long, though certain of them reached 1.5 cm. Addition of HfO_2 also sharply increased
the yield of lamellar crystals. Upon the addition of SiO_2, the main bulk of the crystals emerged in the form of
large nodules of crystals of cubic shape, with the long edge in individual cases reaching 0.4—0.5 cm. Isolated
lamellar crystals formed but rarely, and only for very high cooling rates. Upon the introduction of GeO_2 impurity into the original melt the main bulk of the crystals grew in imperfect cubic shapes, and some in the
shape of prisms with hexagonal or pentagonal cross section; we were only able to obtain lamellar crystals, however, as individuals with edges up to 1 cm on supplementing the GeO_2 in the original melt by $BaCO_3$ in the
stoichiometric ratio necessary to form $BaGeO_3$, using high cooling rates. Impurities of SnO_2 increased the yield

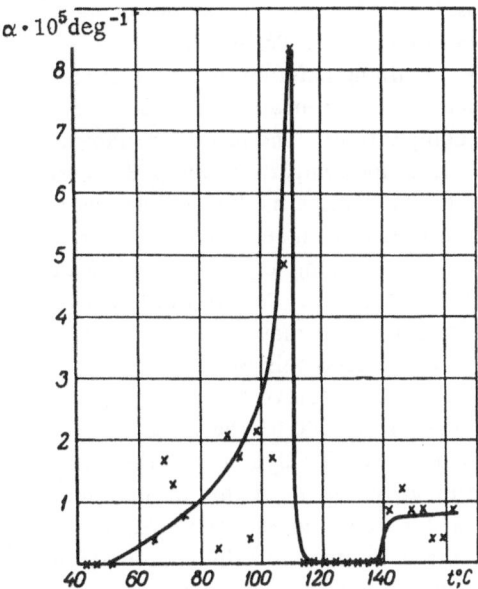

Fig. 2. Temperature-dependence of the coefficient of linear expansion for a crystal containing Ge.

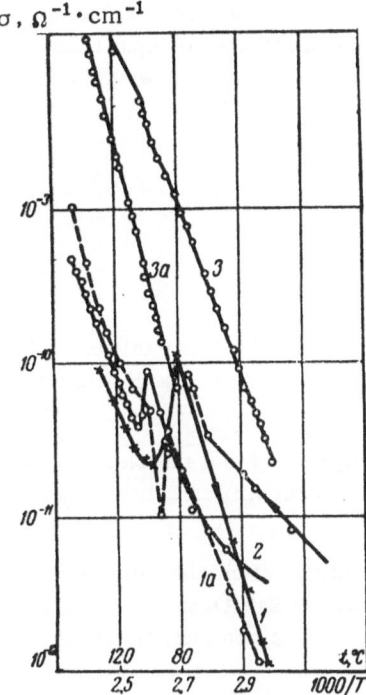

Fig. 3. Tempeature-dependence of electrical conductivity. 1), 1a) for a crystal containing Ge (respectively with silver and aquadag electrodes; 2) Si; 3), 3a) Th (silver electrodes).

of single crystals in the form of lamellar twins, although to a rather smaller extent than HfO_2 and ZrO_2, but together with these formed crystals of cubic shape. The well-faced $BaTiO_3$ crystals with different additives had a clearly expressed domain structure, those with HfO_2 impurity being in general distinguished by a more ordered disposition of the domains.

During the study of the physical properties of the single crystals, it was interesting to examine the possibility of increasing the nonlinearity also as well as the region in which high non-linearity was retained.

By preliminary examination and measurements of a large number of lamellar crystals it was established [17] that the greatest nonlinearity arose in crystals with a smooth domain pattern in the form of narrow, parallel bands [18]. The domain pattern of such crystals was conserved if the crystals were heated above the Curie point and then cooled to room temperature. As we know, the non-linearity of ferroelectrics increases as the phase transformation point is approached. We measured the slope of ε at a frequency of 50 kc/sec for various temperatures. At 80° the dielectric constant at the maximum was twice as great as at room temperature. The coercive force fell in the meantime. Measurements were made on an oscillograph system with respect to the charge loop. Simultaneously, the variation of current was observed through the specimen with field (current loop). Examination of the tempera-ture-dependence of the current loops showed that on heating, to-gether with a sharp peak (rectangular charge loop), a flatter maxi-mum developed, corresponding to a greater field strength. At 90−100°C the sharp peak vanished and hence the process of polarization reversal slowed down. On further heating to the Curie point this new maximum split and the ordinary double charge loops developed.

In the temperature range mentioned, a sharp growth was observed in the linear expansion coefficient $\alpha = (\Delta l / l)(1/\Delta t)$ (Fig. 2) (measurements made on a dilatometer) and large mechan-ical stresses also appeared.

The temperature/conductivity curves of $BaTiO_3$ crystals with impurities also had an anomaly, apparently caused by a sharp jump in polarization [17]. The conductivity was studied for all compo-sitions. The order and character of conductivity was the same for all the crystals, except those containing thorium, in which the con-ductivity was an order higher and had a large temperature hysteresis. The hysteresis increased with increasing electric field strength in the sample. Hence thorium worsens the dielectric properties of the samples, and it is inappropriate to grow crystals with this additive. The ThO_2-contaminated $BaTiO_3$ crystals studied had the form of regularly bounded cubes with 0.3 cm sides. We noticed that the electrodes had an effect on the behavior of the lg δ curve (Fig. 3, curves 1 and 1a). This question, however, requires further

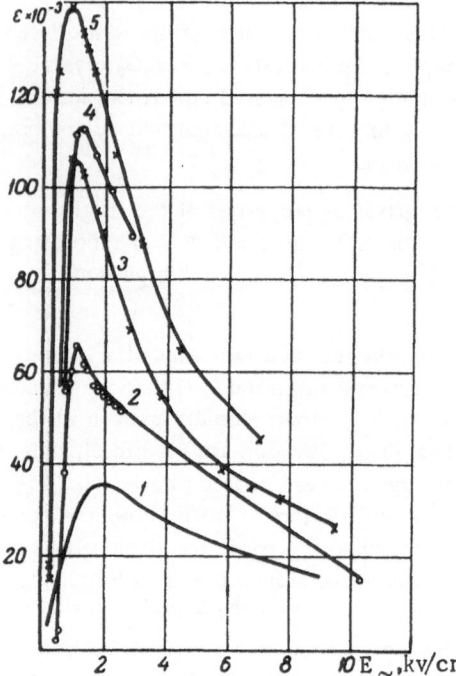

Fig. 4. Variation of ε with alternating
electric field at frequency 50 cps.

examination. We must consider the nonlinear properties of the
crystals with the compositions mentioned. Earlier, the high dielectric constant in crystals containing hafnium for small fields
was indicated [15]. We made analogous measurements at a frequency of 50 cps for other compositions. The results are shown
in Fig. 4. The same figure gives, for comparison, the ε(E~) curve
of ceramic BaTiO$_3$ obtained by the oxalate method and having a
considerably greater nonlinearity than commercial BaTiO$_3$ ceramic.
As seen from the figure, crystals containing any of the various impurities introduced have high nonlinearity. Certain peculiarities
in crystals containing different impurities may, however, be noted.
Crystals containing hafnium were studied only in the form of
plates. All these had high nonlinearity (curve 5) and ε$_{max}$ at
frequency 50 cps reached values of 200,000 for individual crystals.
Crystals with germanium impurity were studied both in the form of
plates and cubes. Good nonlinearity was found only in the lamellar
crystals (curve 4). Crystals with silicon impurity were examined as
both plates and cubes. In contrast to those containing germanium,
these crystals also included cubic forms with large nonlinearities
(curve 3). Crystals of this composition, however, showed large
fluctuations in properties, since the domain structure in crystals of
cubic form was unstable. Having taken measurements at a frequency of 1 kc/sec, we noted that crystals containing Si had the
lowest dispersion; whereas the ratio of the maximum values ε$_{max}$
at 50 cps and ε$_{max}$ at 1 kc/sec was 2 for other compositions, for
this composition it did not exceed 1.5. Curve 2 shows the ε(E~)
relation for a crystal containing Zr.

The reversible nonlinearity N$_r$ of the crystals was also determined:

$$N_p = \frac{1}{\varepsilon} \frac{\Delta\varepsilon}{\Delta E_=}$$

at a frequency of 1 kc/sec for various alternating electric field strengths. The main bulk of crystals containing
Si were individual well-faced crystals in which N$_r$ reached 6 or 7. For crystals containing Sn the N$_r$ did not
exceed 0.9; for those containing Ge, N$_r$ did not exceed 2.5.

Upon the measurement of the piezoelectric properties of the crystals, it was found that the piezo modulus
had a sharp maximum near the phase transformation point. Crystals without mechanical stresses had large piezo
moduli.

All these results lead to the conclusion that the introduction of small quantities of dioxides of elements
from the fourth group of the period table into the original melt, during the growth of BaTiO$_3$ single crystals
from a solution in potassium fluoride using the same condition of crystallization, has a considerable effect on
the form and yield of the crystals formed.

The most perfect crystals, with a large yield of lamellar shapes, are formed upon the introduction of
halfnium and zirconium dioxides into the original melt. Barium titanate single crystals containing very small
quantities of these elements (up to 0.1%) have high nonlinear and piezoelectric properties. The parameters
of the single crystals measured (dielectric constant, linear expansion coefficient, electrical conductivity, and

piezo modulus) are distinguished by a sharp jump at the phase transformation point. In view of the sharp rise in the linear expansion coefficient near the phase transformation, which produces mechanical stresses and worsens the nonlinear properties of the single crystals, the crystals for use as nonlinear elements require temperatures of 90 to 110°C.

Literature Cited

1. A. L. Khodakov and M. L. Sholokhovich, Dokl. Akad. Nauk SSSR, 141(2):338 (1961).
2. M. L. Sholokhovich and A. L. Khodakov, Vopr. Radioelektron., III–Vol. 4; 108 (1961).
3. R. C. Devries, Am. Ceram. Soc., 42(11):(1959).
4. Atuo Nishioka, Kyozo Sekikawa, and Massakazu Owaki, J. Phys. Soc. Japan, 11(2):180 (1959).
5. C. Pulvari, J. Am. Ceram. Soc., 42(8): 355 (1959).
6. L. M. Berberova, Abstracts of the Fourth Scientific Conference of Graduates of Rostov State University (Izd. RGU, 1962).
7. G. A. Smolenskii, M. A. Karamyshev, and N. I. Rozgachev, Dokl. Akad. Nauk SSSR, 79:53 (1951).
8. G. A. Smolenskii and V. A. Isupov, and Akad. Nauk SSSR, 96(1):58 (1954).
9. G. A. Smolenskii, Zh. Neorgan. Khim., 1:1402 (1956).
10. T. N. Verbitskaya, Technology of Preparing Variconds and Their Properties (Press of the Moscow House of Scientific–Technical Propaganda, 1951).
11. M. L. Sholokhovich, and A. L. Khodakov, Growth of Crystals Vol III (Moscow, 1961), p. 463.
12. T. N. Verbitskaya, Elektrichestvo, 8:68 (1960).
13. T. N. Verbitskaya, E. I. Gindin, and V. G. Prokhvatilov, Collection: Solid State Physics, Vol. I (Izd. Akad. Nauk SSSR, 1959), p. 180.
14. T. N. Burakova and T. N. Verbitskaya, Dokl. Akad. Nauk SSSR, 126(5):994 (1959).
15. O. I. Prokopalo and E. G. Fesenko, Collection: Ferroelectrics (Izd. RGU 1961).
16. M. L. Sholokhovich, A. L. Khodakov, and T. N. Lezintseva, Crystallization and Phase Transformations (Izd. Akad. Nauk Belorus. SSR, 1962), pp. 426-37.
17. T. N. Lezqintseva, Abstracts of the Fourth Scientific Conference of Graduates of Rostov State University (Izd. RGU, 1962), p. 72.
18. M. L. Sholokhovich, A. L. Khodakov, T. N. Lezintseva, and V. I. Varicheva, Collection: Ferroelectrics (Izd. RGU, 1961), p. 12.

STUDY OF THE VELOCITY OF PROPAGATION OF ULTRASOUND
IN TRIGLYCINE SELENATE IN THE CURIE TEMPERATURE REGION

N. N. Sirota, N. P. Tekhanovich, and V. M. Varikash

In 1957 Matthias, Miller and Remeika discovered the ferroelectric properties of triglycine sulfate and its isomorph, triglycine selenate [1]. Since then, various properties of triglycine sulfate, including the change in the propagation velocities of ultrasound and elastic moduli at the Curie point, have been studied by many workers [2–4]. Up to the present, however, hardly any papers exist on the study of triglycine selenate.

The purpose of the present investigation is to study the variation in the propagation velocity of elastic ultrasonic waves in the Curie temperature interval in different crystallographical directions, this being a continuation of an examination of the changes in the ultrasound propagation velocities and elastic properties of ferroelectrics and ferromagnetics at the Curie point [4, 5].

A study of the propagation velocities of transverse and longitudinal ultrasonic oscillations in triglycine selenate was made with the UZIS-7 system at a frequency of 5 Mc/sec in the temperature range $^-20-^+50°C$.

The method of determining the ultrasonic propagation velocities was described earlier in [4].

The triglycine selenate was obtained by the action of selenic acid on a solution of glycine. Single crystals were grown from supersaturated solutions by temperature reduction. Samples in the form of cubes with 10 to 14 mm sides were cut from a large triglycine selenate crystal for examination.

In all the samples, opposite faces were kept parallel, to an accuracy of 0.01 mm. Dry ice was used to obtain negative temperatures during the investigation. Ethyl alcohol was used as the thermostating liquid.

Figure 1 shows curves of the propagation velocity of longitudinal ultrasonic waves as a function of temperature for three crystallographical directions, [100], [010], and [001]. As the temperature rises to the Curie point, the ultrasound propagation velocity in all three directions falls uniformly. In the Curie temperature interval the velocities pass through a minimum. The largest anomaly of the velocity in this region occurs for the crystallographical direction [100].

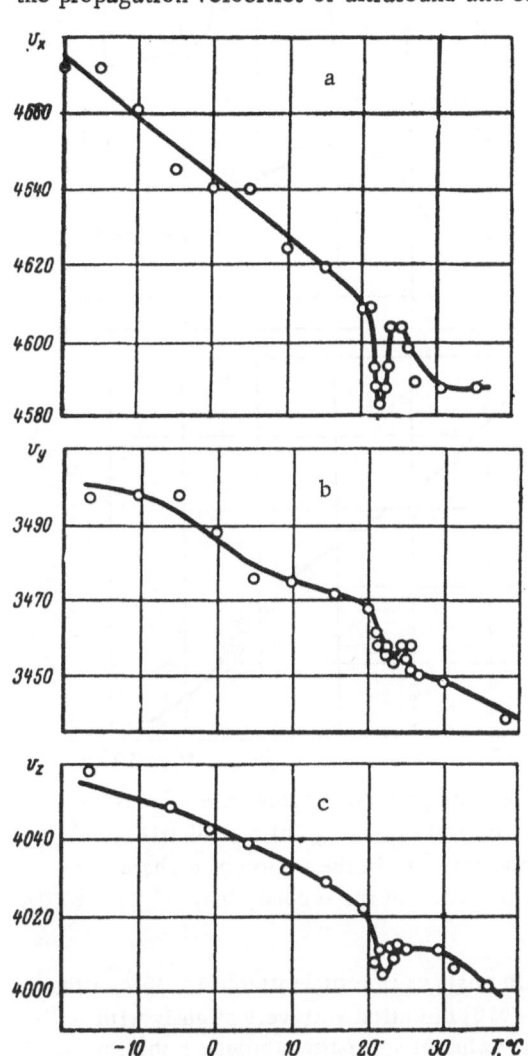

Fig. 1. Temperature-dependence of the propagation velocity of longitudinal ultrasonic waves in triglycine selenate in the directions a) [100], b) [010], and c) [001].

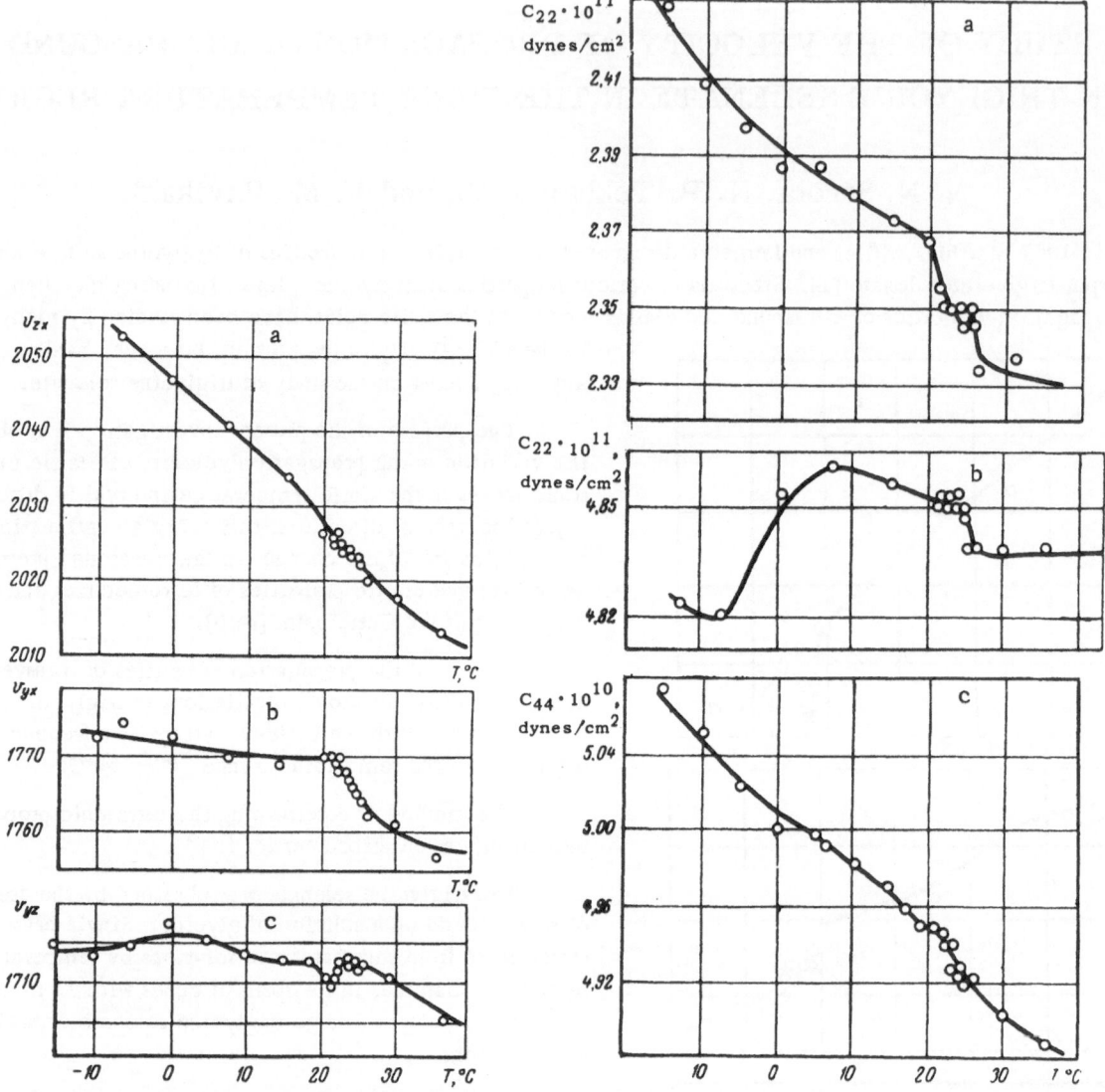

Fig. 2. Temperature-dependence of the propagation velocity of transverse ultrasonic waves in triglycine selenate in the directions a) [001], b), c) [010].

Fig. 3. Temperature-dependence of elastic moduli a) c_{22}, b) c_{44}, c) c_{66} of triglycine selenate in the region of a phase transformation of the second kind.

The propagation velocity of transverse waves varies little with changes in temperature (Fig. 2). As the figure shows, the velocities v_{yx} and v_{yz} of transverse waves in the [010] direction vary very slightly with temperature, except that in the region of the Curie temperature the value of v_{yz} passes through a shallow minimum and that of v_{yx} falls sharply in this temperature range. The velocity v_{zx} of the transverse wave in the [001] direction, however, falls almost linearly over the entire temperature range. From the measured propagation velocities of ultrasound [6], the elastic moduli c_{22}, c_{44}, and c_{66} were calculated. Figure 3 shows the temperature-dependence of these elastic moduli.

The variation of the elastic moduli with temperature measured for triglycine selenate shows a certain correspondence with that measured for triglycine sulfate. In triglycine selenate, however, the anomalies in the ultrasound propagation velocities and elastic moduli at the Curie point are considerably less evident than in triglycine sulfate. Perhaps this is connected with the smaller value of the Curie point in triglycine selenate. The variation in Bragg reflection intensities with temperature near the phase transformation is larger for triglycine selenate than for triglycine sulfate.

The problem of the anomaly in the propagation velocities of ultrasound may find a satisfactory explanation in the general thermodynamic theory of phase transformations of the second kind [7–9].

Literature Cited

1. B. T. Matthias, C. E. Miller, and I. P. Remeika, Phys. Rev., 104:849 (1956).
2. V. P. Konstantinova, I. M. Sil'vestrova, and K. S. Aleksandrov, Collection: Physics of Dielectrics (Moscow, 1960), pp. 351–65.
3. L. A. Shuvalov and K. A. Pluzhnikov, Kristallografiya, 6(5):692 (1961).
4. N. N. Sirota, V. M. Varikash, and N. P. Tekhanovich, Collection: Crystallization and Phase Transformations, Izd. Akad. Nauk Belorus. SSR (1962), pp. 435–439.
5. N. N. Sirota, E. A. Ovseichuk, and N. P. Tekhanovich, Collection: Ferrites (Izd. Akad. Nauk Belorus. SSR, Minsk, 1960). pp. 74–77.
6. K. S. Aleksandrov, Kristallografiya, 3(5): (1958).
7. L. D. Landau, Zh. Eksp. i Teor. Fiz., 7(1):(1937).
8. V. K. Semenchenko, Zh. Fiz. Khim., 35(11): (1961); Collection: Application of Ultraacoustics to the Study of Matter, Vol. III (MOPI, Moscow, 1956).
9. M. A. Leontovich, Introduction to Thermodynamics (GITTL, Moscow, 1959).

STUDY OF X-RAY SCATTERING IN CRYSTALS OF SOME
FERROELECTRICS IN THE REGION OF THE CURIE TEMPERATURE

N. N. Sirota, V. M. Varikash, and É. A. Ovseichuk

The study of X-ray scattering in the neighborhood of the Curie point of ferroelectrics has attracted the attention of scientists because the experimental data thus obtained may provide information as to the precise nature of the ferroelectric transformation and its mechanism.

In 1937, in the first papers to appear devoted to the theory of phase transformations of the second kind, Landau considered the possible variation in the intensity of X rays scattered by a crystal in the neighborhood of a phase transformation of the second kind [1]. Landau showed that, near the Curie point, there should be a rise in the incoherent scattering intensity and a sharp fall in that of the coherent scattering, owing to an increase in fluctuations and the dynamic displacement of ions from equilibrium positions.

The phenomenon of anomalous neutron and X-ray scattering near a ferromagnetic transformation of the second kind was observed by Umanskii and Veksler [2], Lowde [3], Riste and MacReynolds [4], Ovseichuk, and Tekhanovich [5], and others. In the region of ferroelectric transformations critical scattering was studied by the present authors [6] as well as by Shibuya and Mitsui [7].

The comparatively small amount of experimental data in existence indicates that the original form of Landau's theory of critical X-ray scattering in the neighborhood of phase transformations of the second kind, although not without significance, is nevertheless insufficient, and in its initial form cannot give a complete explanation of all the anomalies observed.

The Landau theory of X-ray scattering near the temperatures of phase transformations of the second kind was further developed by a number of authors, notably Krivoglaz [8], who also proposed a phenomenological explanation which neglected the specific structure of the crystals. Hence the comparision of theoretical results with experimental data in crystals of low symmetry, such as triglycine sulfate, meets with considerable difficulty.

Systematic investigations of critical X-ray scattering in the region of ferromagnetic and ferroelectric transformations were undertaken in our laboratory. The present paper presents the main experimental results on X-ray scattering near the ferroelectric transformation in triglycine sulfate, triglycine selenate, and Rochelle salt.

The methods used for carrying out these studies were practically identical. The samples for investigation were obtained in the form of a power finely ground in an agate mortar, annealed for an hour after grinding. The X-ray recordings were taken in the URS-50I X-ray system with a Geiger—Müller counter in copper K_α radiation. The samples were fixed to the table of the goniometer in a specially constructed thermostat, the temperature of which was maintained by means of a flow of constant temperature water from an ultrathermostat or by means of an electric heater. While the recordings were taken the sample temperature was kept constant to 1°(in some cases, working with the ultrathermostat, the temperature regulation was considerably more precise). Three X-ray recordings were taken at each set temperature. The entire heating cycle was duplicated and also repeated three times. The line intensities were determined by planimetering the X-ray records obtained on the automatic recorder EPP-09. Figure 1 shows the intensity of individual lines of the Debye pattern of triglycine sulfate as a function of temperature. We notice the different way in which the intensities of different X-ray lines vary with temperature. The intensity of line (031) (Fig. 1a) rises as the

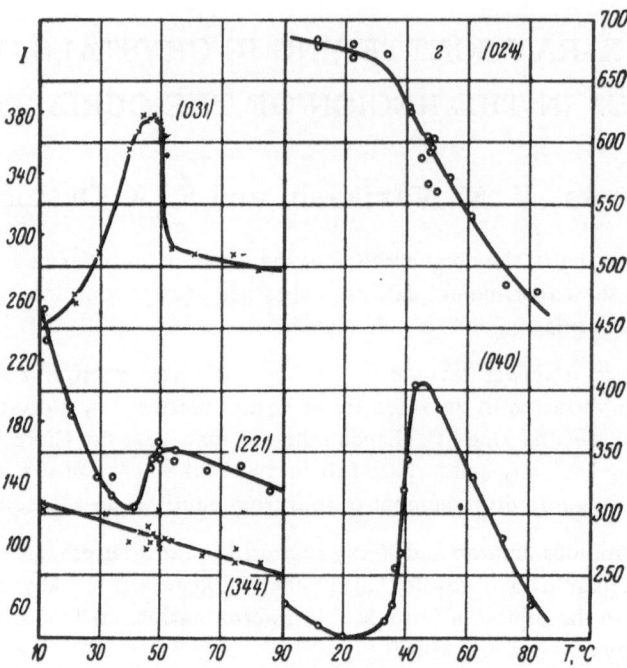

Fig. 1. Temperature-dependence of X-ray line intensities
of triglycine selenate (in arbitrary units).

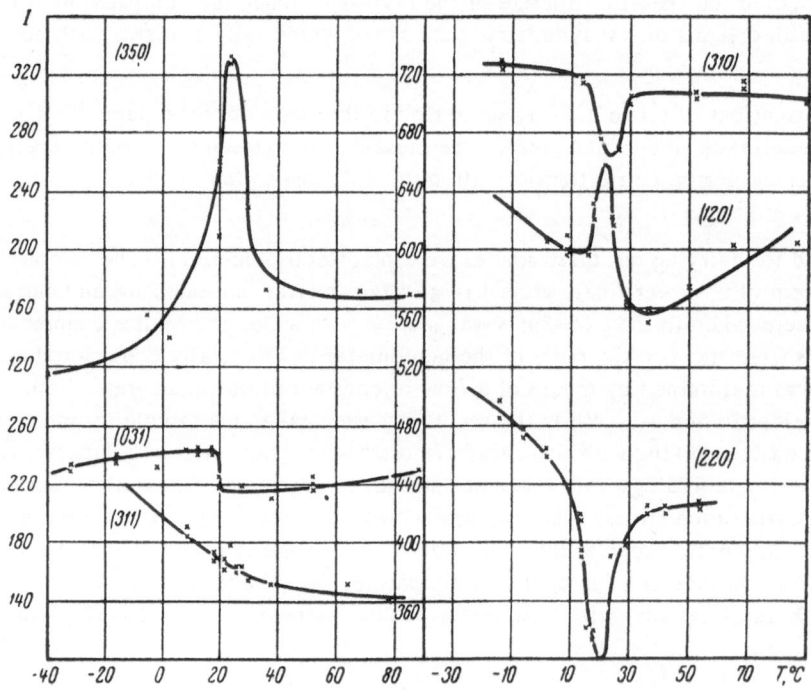

Fig. 2. Temperature-dependence of X-ray line intensities of triglycine
selenate (in arbitrary units).

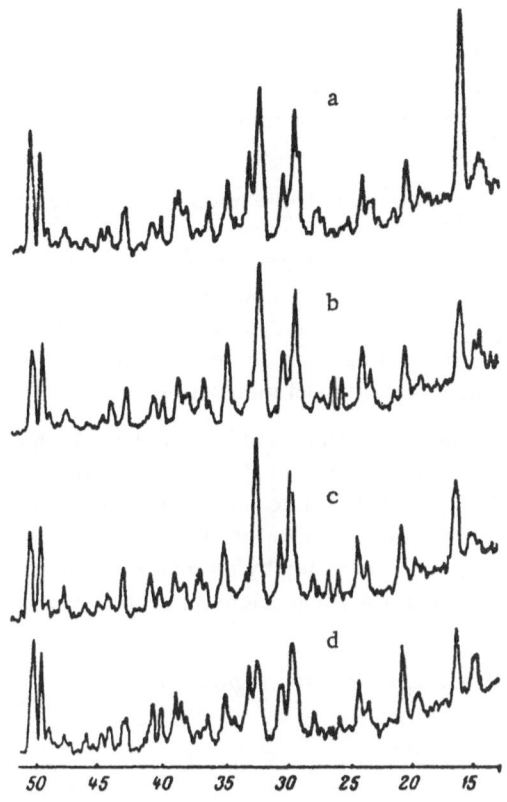

temperature increases up to the Curie point, reaches a maximum near the Curie temperature, and then falls sharply on further heating in the range 50—55°. Above 55° there is a gradual fall in intensity. The intensity of the (221) line (**Fig. 1b**), on the other hand, falls as the temperature rises from 10 to 40°, reaches a minimum at 40°, and rises from there to the Curie point. After reaching the Curie point there is a gradual fall in intensity. In the (024) and (344) lines (Fig. 1d, c), the intensity/temperature relation is different. The intensities of these lines fall smoothly as the temperature rises from 10 to 80°C, without any obvious anomalies in the neighborhood of the Curie point. It may be that there are insignificant anomalies characterized by deviations from the smooth curves within the limits of experimental error. This idea is supported by a somewhat greater scatter of experimental points near the Curie temperature.

Figure 2 shows curves for the intensity variation of the (120), (220), (310), (031), (311), (350) X-ray reflections of triglycine selenate. The intensity variation of the (350) and (120) lines with temperature have a certain similarity. Maximum intensity is observed near the Curie temperature. The (220) and (310) lines, however, have a sharp intensity minimum near the Curie point. The intensity of the (031) line rises as temperature increases from -50° to the Curie point. At the Curie point there is a sudden drop, after which the intensity rises slightly as the temperature increases to 90°C. The intensity of the (311) line falls smoothly with increasing temperature, with no anomaly near the Curie temperature.

Figure 3 shows X-ray records of Rochelle salt taken at various temperatures. As seen from the figure among the large

Fig. 3. Debye power record of Rochelle salt at various temperatures. a) 8°C, b) 20, c) 26.5, d) 44°C.

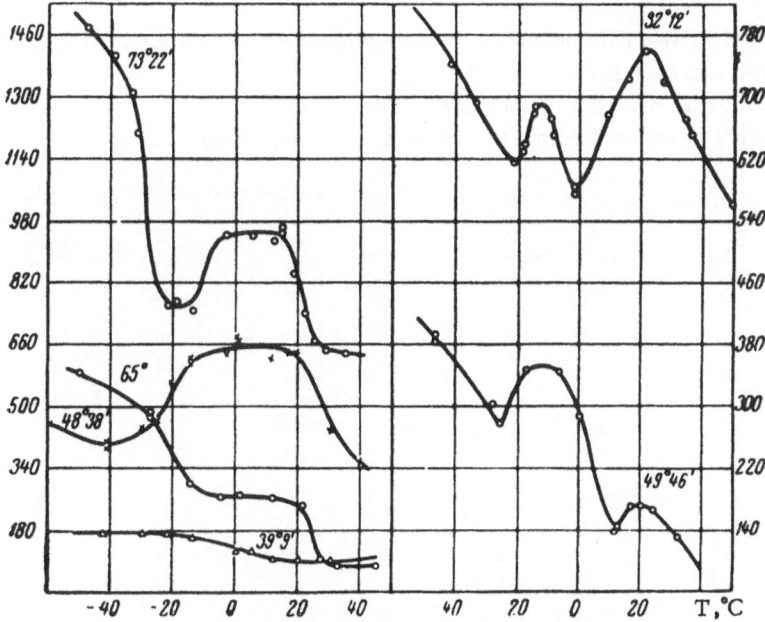

Fig. 4. Temperature-dependence of the X-ray line intensities of Rochelle salt.

number of lines we can separate out some for which the intensity changes considerably near the Curie temperature, and others for which the intensity variation is slight.

Figure 4 shows the intensity variation curves on heating from - 50° to + 50°C for comparatively temperature-sensitive X-ray lines. As we know, Rochelle salt has two Curie temperatures: the lower is at - 18°C and the upper at + 24°C. Thus our temperature range embraces both the region in which the ferroelectric state exists and the higher temperature regions.

The variation in line intensity with temperature, and in particular that near the Curie temperature, may be divided into several groups of similar curves. For example, reflections with reflection angles $2\Theta = 73°22'$, $65°$; $48°38'$ are characterized by a fall in intensity on heating from - 40 to 0°C, and a maximum or flat part of the curve in the region where the ferroelectric state exists. The reflection with $2\Theta = 39°9'$, however, has no maximum or flat part at all on the intensity/temperature curve in this region. The intensity/temperature curves for lines with $2\Theta = 32°12'$ and $49°46'$ are characterized by two maxima, corresponding to the lower and upper Curie points respectively. We may note the difference in the intensities at these maxima and their ratios.

The data thus obtained indicate the significance of the observed intensity variation effects in the neighborhood of the Curie point. Our experimental study enables us to draw conclusions regarding the physical and crystallochemical nature of ferroelectric transformations in the neighborhood of the Curie temperature, characterized by a change in the mean square dynamic displacements of ions from their equilibrium positions, and also, apparently, their positions in the lattice (degree of order). This is particularly indicated by the complete disappearance of certain Debye lines at the Curie point in triglycine sulfate and triglycine selenate, as well as in Rochelle salt.

Further X-ray and neutron diffraction studies of this problem will undoubtedly arouse great interest.

Literature Cited

1. L. D. Landau, Zh. Eksper. i Teor. Fiz. 7(11):(1937).
2. V. P. Veksler and Ya. S. Umanskiii, Sov. Phys., 6:258 (1934).
3. R. D. Lowde, Rev. Mod. Phys., 30(1):69–74, (1958).
4. A. W. MacReynolds and T. Riste, Phys. Rev., 95:1161 (1954).
5. N. N. Sirota, É. A. Ovseichuk, and N. P. Tekhanovich, Collection: Ferrites (Izd. Akad. Nauk Belorus. SSR, Minsk, 1960), pp. 74–77
6. N. N. Sirota and V. M. Varikash, Contribution to the Symposium on Ferroelectricity and Ferromagnetism. Summaries of Contributions (Leningrad, May 1963).
7. I. Shibuya and T. J. Mitsui, Phys. Soc. Japan, 16(3):479 (1961).
8. M. A. Krivoglaz, Theory of Ordered Alloys (Fizmatgiz, Moscow, 1958).

PART TWO

SOLID STATE TRANSFORMATIONS

THERMAL EFFECTS OF THE TRANSFORMATIONS OF METALS
AND SEMICONDUCTORS IN THE SOLID AND LIQUID STATES

N. A. Nedumov and V. K. Grigorovich

The fusion of a crystalline substance differs from polymorphic transformations in the solid state in that it involves the disruption of long-range order while (in most cases) preserving the short-range order characterizing the melting crystal; this happens after certain critical concentration of vacancies is reached (a steady state in which a definite proportion of the interatomic bonds are broken), and corresponds to a considerable thermal effect. Polymorphic transformations have a different nature, in that they are accompanied only by a temporary disruption of the bonds between atoms at the time of the actual transformation, after which the interatomic bonds are restored; this connates a change rather than a disruption of long-range order. We may suppose that a change in crystal structure depends to a greater extent than does melting on the electronic structure of the atoms. In view of this, it is interesting to study the preparation of a crystal structure for a phase transformation on approaching the transformation temperature, the completion of the transformation in the region where the new phase exists, the hysteresis of the melting and polymorphic transformation processes on cooling, and so on. There is also some point in studying thermal effects which might indicate changes of short-range order in the liquid state, analogous to the thermal effects of polymorphic transformations in the solid state. This might happen, for example, when the melt of a polymorphic substance is subject to severe supercooling.

For a comparative study, we chose four pure transitional metals, — manganese, iron, cobalt, and nickel — covalent—metallic and covalent elements of the principal groups — gallium, indium, silicon, germanium, antimony, and bismuth — and two semiconducting compounds with the zinc blende structure, namely, gallium and indium antimonides. The specification of the materials studied is given in Table 1. The electrolytic manganese contained 99.83% Mn. The electrolytic iron, refined in hydrogen and vacuum-remelted, contained 99.99% Fe. The cobalt had a slight nickel impurity (1.5%), and the nickel a little cobalt (0.4%). Both metals also contained traces of iron. The remaining substances were of a very high degree of purity, characteristic of materials used for producing semiconductors.

The investigation was made in a noncontact thermal analysis system [1].

The portions of material for investigation indicated in Table 2 were weighed on semimicroanalytical scales to an accuracy of 0.00001 g and placed in an alumina crucible of dimensions d = 9 mm, h = 12 mm with wall thickness ~0.8 mm, or in a beryllium oxide crucible with d = 9 mm, h = 11 mm, and a wall thickness of 0.5 to 0.8 mm. The weight of the alumina crucible was ~2.12 g, and that of the beryllium oxide crucible 0.95 g. After placing the sample in the furnace, this was evacuated to 10^{-4} or 10^{-5} mm Hg and heated to 400 or 500°C to remove occluded gases, after which it was filled with helium, carefully purified from water and gases by passing through silica gel and activated charcoal cooled to - 196°C by liquid nitrogen. Heating was carried out at 40—70 deg/min (second rate) or 17—35 deg/min (third rate). Cooling took place at 25—45 or 10—22 deg/min.

After the experiment, the crucible with specimen was weighed and the weight loss was determined. There was only a significant evaporation of metal in the case of manganese (1.1%). Evaporation of gallium, indium, silicon, and germanium was slight. Gallium and indium antimonides evaporated more markedly. but even for three or four times repeated heatings to 1000°C the weight loss was no more than 3.3% for GaSb and 0.7% for InSb. Antimony evaporated quite intensely on heating to 800 or 900°C. Bismuth did so less intensely.

TABLE 1. Original Materials Used in the Investigation

Material	Specification	Purity, %	Impurities
Manganese	Electrolytic	99,83	trace S
Iron	Electrolytic, refined in hydrogen	99,99	$O_2 < 0,001\%$ $Cu—0,001\%$
Cobalt	Electrolytic	98,00	$Ni—1,5\%$ $Fe—0,5\%$
Nickel	Granulated	99,5	$Co—0,4\%$ $Fe—0,1\%$
Gallium	Semiconductor purity	99,999	$Al < 1 \cdot 10^{-4}$ $In < 3 \cdot 10^{-3}$ $Cu—3 \cdot 10^{-5}$ impurities 10^{-4}
Indium	Semiconductor purity	99,99	Σ impurities 10^{-3}
Silicon	Commercial, semiconductor	99,9999	Σ impurities 10^{-4}
Germanium	semiconductor purity	99,99999	Σ impurities $10^{-5}—10^{-6}$
Antimony	Commercial, Semiconductor purity	99,999	Σ impurities 10^{-3}
Bismuth	Semiconductor purity	99,99	Σ impurities 10^{-2}
GaSb	Semiconductor purity	99,99	Σ impurities 10^{-2}
InSb	Semiconductor purity	99,99	Σ impurities 10^{-2}

Heats of fusion of the substances studied were determined from the areas of the maxima bounded by the differential record. The thermal effects were measured on a calibrated scale of thermal susceptibility $q_s = Q/s = F(\overline{T}, P, \overline{V}, \Sigma\alpha)$. The scale was constructed as a function of temperature T and the total heat emission of the block system $\Sigma\alpha/V$ for constant gas pressure P and heating rate $\overline{V}$. Here Q denotes the thermal effect of the process and S the area bounded by the differential curve. The results are presented in Figs. 1—5 and Table 3. In manganese (Fig. 1), thermal effects corresponding to the following transformations were observed on heating: $\alpha \rightarrow \beta$ at 727° (725 to 730°), $\beta \rightarrow \gamma$ at 1094° (1085 to 1095°), $\gamma \rightarrow \delta$ at 1134° (1130 to 1135°) and melting at 1244° (1240—1244°), which agreed with published data. Fusion was completed rapidly, and in the liquid state thermal effects were not observed. Solidification took place without marked supercooling at 1240 to 1245°. The transformations of δ−Mn (centered cubic structure) into λ−Mn (dense cubic packing) and γ−Mn into β−Mn (composite cubic structure) took place without appreciable hysteresis at 1130—1135° and 1070—1094°C, respectively. The transformations of composite cubic structure β−Mn into composite cubic structure α−Mn, however, took place with a large hysteresis, reaching 100 to 182°. The absence of supercooling on solidification is probably due to insufficient purity of the samples investigated (99.87% Mn).

Upon the heating of pure iron (Fig. 1), thermal effects appear at the magnetic $\alpha \to \beta$ —transformation (770 to 780°C), the $\beta \to \gamma$ —transformation (910°), the $\gamma \to \delta$ —transformation (1400 to 1405°), and on melting (1535—1540°). The melting process is completed in a short time. Upon the cooling of relatively impure iron, hardly any hysteresis of solidification or polymorphic transformations appear. After melting in purified helium, pure iron tends toward severe supercooling. Solidification of the iron is observed at 1435, 1425, 1380, and 1350° and even at lower temperatures (1230°). In the liquid state, thermal effects are observed on cooling at 1540—1490°. The temperature of the $\delta \to \gamma$ —transformation on cooling lies between 1390 and 1395°. In two recent cases it was not recorded at all; after solidification only the $\gamma \to \beta$ and the magnetic transformation were found, the first taking place with some hysteresis (860—907°), and the second at 760°. The lower the temperature at which solidification took place, the more sharply expressed was the thermal effect in the liquid iron cooling section. Supercooling of iron in alumina crucibles reached 250—310°, and in beryllium oxide crucibles 200°,

TABLE 2. Experimental Conditions of Thermograph Recording

Material	No. of experiment	Weight, g		Crucible material	Rate, deg/min		Max. temp., °C
		Original and final	Weight loss,%		heating	cooling	
Manganese	1	2.29		Al$_2$O$_3$	57	37	1450
	2			»	57	37	1450
	3	2.25	1.1	»	57	37	1450
Iron	1	4.72		Al$_2$O$_3$	41	36	1650
	2			»	42	30	1650
	3	4.6914	0.5	»	55	42	1650
	4	4.09	0.5	»	42	42	1650
Cobalt	1	2.7265		BeO	46	37	1600
	2	2.7172	0,05	»	46	37	1600
	3	2.7300	0.1	»	46	37	1600
Nickel	1	2.6228		»	40	30	1600
	2	2.5181	0.002	»	39	29	1600
Gallium	1	1.2584	0,003	BeO	35		970
Indium	1	2.2018	0.00002	»	17	10	600
	2	2.20745	0.01	»	68	48	1515
Silicon	1	0.4442		BeO	35	34	1655
	2	0.4424	0,2	»	40	36	1755
Germanium	1	1.49727		BeO	55	40	1260
	2			»	55	40	1260
	3	1.49344	0,003	»	55	45	1260
Antimony	1	1.70725	30	BeO	62	47	1835
	2	1.4941		»	34	22	1610
	3	1.1365	24	»	34	22	1610
Bismuth	1	2.0085		BeO	74	40	1595
	2	1.25319	27		60	46	1680
GaSb	1	0.871		BeO	26	21	1050
	2			»	26	21	1050
	3	0.84236	0.3	»	26	21	1050
InSb	1	1.2395		BeO	23	21	1000
	2	0.23078	0,008	»	23	21	1000

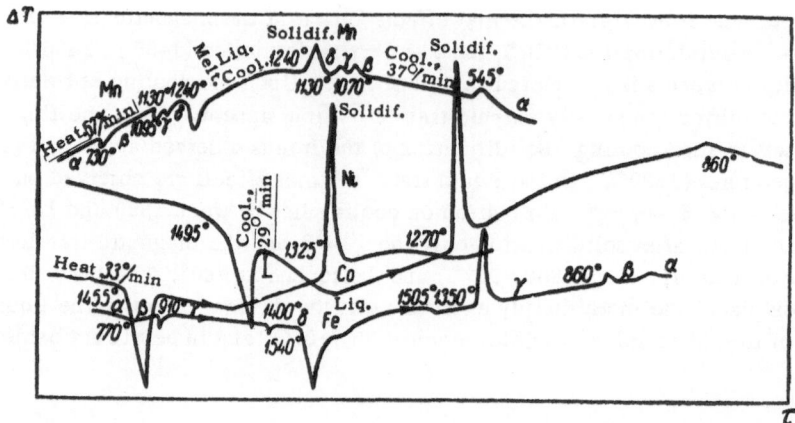

Fig. 1. Thermal effects in the phase transformations of manganese, iron, cobalt, and nickel.

depending little on the cooling rates between 30–50 and 10–20 deg/min. When solidification occurred below the $\delta - \gamma$ – transformation temperature, the thermal effect corresponding to this transformation did not appear. Apparently molten iron, having a short-range order corresponding to the centered cubic structure of δ–Fe, can solidify directly into the γ–phase, i.e., solidification can coincide with the polymorphic $\delta - \gamma$ – transformation, involving a change of short-range order. In these cases, no sign of thermal effects between solidification and the $\gamma \rightarrow \beta$ –transformation was found, and the latter transformation took place with comparatively little hysteresis at 900–845°.

Phase transformations in iron on heating and cooling began suddenly without noticeable preparation of the structure, and ended without hysteresis. The change in heat capacity with the magnetic $\alpha \rightleftharpoons \beta$–transformation began long before 768° was reached, at which point it reached a maximum. The heat of fusion of iron was 3.686 kcal/g. atom.

On heating cobalt, an $\alpha \rightarrow \beta$ –transformation at 355 to 390°, a magnetic transformation at 1100–1095°, and fusion at 1495–1500° were observed (see Fig. 1). In the molten state any thermal effects which might indicate a change in short-range order were unobserved. Solidification took place at 1365, 1340, and 1270°, i.e., with hysteresis at 135–230°. Upon the heating of nickel (See Fig. 1) there was a magnetic transformation at 358° and fusion at 1440–1455°. Thermal effects which might have indicated a change in coordination number in the liquid state were not observed. Solidification occurred at 1375–1325°, i.e., with supercooling at 80–130°. Melting and solidification in cobalt and nickel took place without noticeable preparation in the original state, as may be judged from the sharp change in the curves on reaching the melting or freezing points. Completion of these phase transformations was not extended in time.

In gallium, melting began without special preparation in the solid state and took place at 30°. Disruption of the short-range order characteristic of the solid state, however, continued after melting up to a temperature of 225° (Fig. 2). Gallium supercooled below 18°, and its solidification could not be recorded.

Indium melted at 155–156°, also without noticeable preparation effects before melting (Fig. 2). Disruption of the short-range order characteristic of the solid state continued in the melt up to at least 180°C. On cooling, the molten indium supercooled to 135–145°. Solidification took place without marked thermal effects in the liquid state, which is in accordance with data indicating the absence of any change in coordination number for indium on melting.

No sharp bend in the curve for silicon was observed on reaching the melting point (Fig. 3); the curve rose very smoothly, beginning at 1350–1380°. This apparently indicates the existence of a preparatory effect before

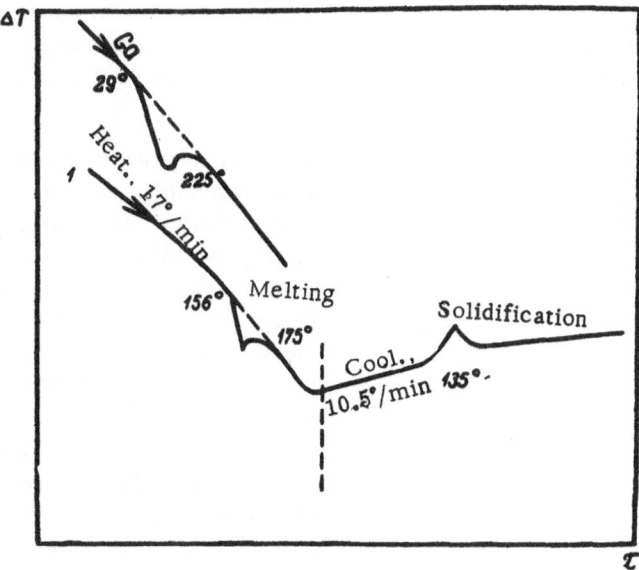

Fig. 2. Thermal effects in the phase transformations of gallium and indium.

melting. Complete disruption of the short-range order characteristic of the solid state occurred above the melting point and ended at 1545—1565°. On cooling, no thermal effects were observed right down to the freezing point, which indicates that molten silicon retains short-range order and properties corresponding to its transformation from the semiconducting state to the metallic state. No marked supercooling of silicon was found. Solidification took place at 1410—1380°, and on further cooling in the solid state no thermal effects were observed. In the more electropositive germanium (Fig. 3), the preparation of the structure before melting was also noticeable in the solid state, although the onset of melting gave a more marked break in the curve. Melting occurred at 935—937°. The disruption of the short-range order peculiar to the solid state was completed on further heating of the liquid to 1010—1100°. Right down to the onset of solidification, cooling was not accompanied by thermal effects, thus indicating the stability of the metallic state in molten germanium. Solidification took place at 775—850°, i.e., with considerable supercooling. In the solid state no thermal effects were observed.

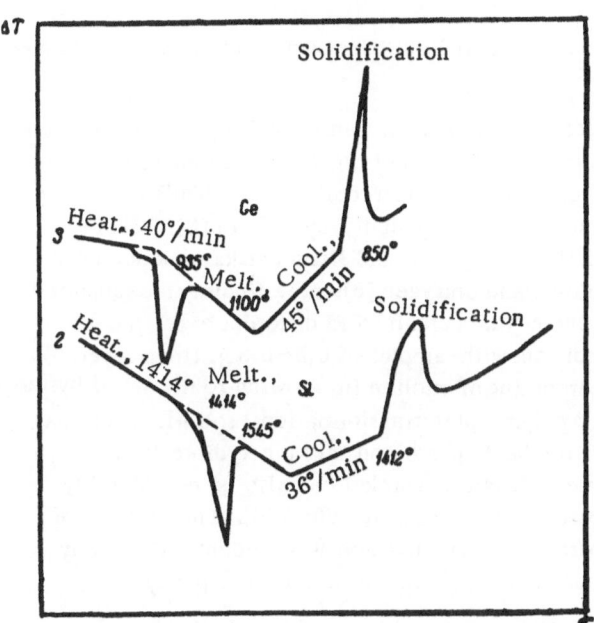

Fig. 3. Thermal effects in the phase transformations of silicon and germanium.

With antimony, just as was the case with silicon, a considerable thermal effect of preparation for melting was observed (Fig. 4); this started at 600°, and there was no sharp bend corresponding to the onset of melting. After the transformation, heat continued to be absorbed right up to 700 or 750° in the liquid state, apparently due to the breaking up of directional bonds. Upon the cooling of antimony, after heating to 710—780°, right down to the onset of solidification, there were no thermal effects such as might indicate a change in short-range order in the liquid state. In other words, molten

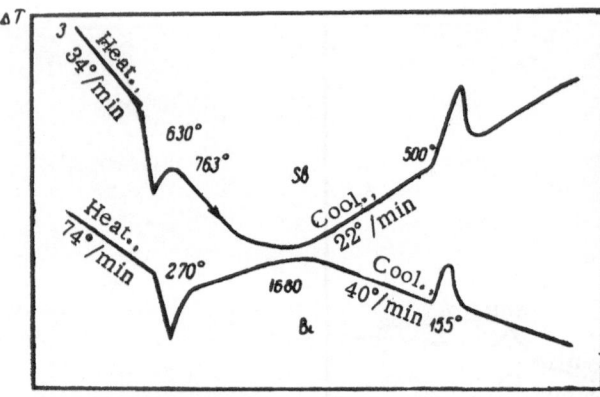

Fig. 4. Thermal effects in the phase transformations
of antimony and bismuth.

antimony firmly maintains the metallic state. The
supercooling of antimony was slight, and solidification
occurred at 620, 610, and 500°C.

In bismuth the process of preparing for melting
in the temperature range close to the melting point
(270°) is less marked than in antimony, and there is
a clear break in the curve marking the onset of melting (Fig. 4). The process of breaking up the directional
bonds extends to approximately 400°. On cooling of
the molten metal after further heating, no thermal
effects were observed in the liquid state right down to
the freezing point. Solidification took place at 235
and 155°, i.e., with supercooling of 35−115°. In the
solid state, no thermal transformation effects were observed down to room temperature. Examination of the
semiconducting compounds GaSb and InSb (Fig. 5)
gave the following results. In gallium antimonide
preparation for melting was observed, starting from
~510−650°. Melting began at 700−710°. The disruption of the covalent bonds, judging by the thermal effects,
ended at about 800°C. After heating the melt to 1050−1080° and cooling to the onset of solidification, there
were no signs of thermal effects. Solidification took place at 615 to 625°, i.e., with a supercooling of ~75−85°.

The melting of indium antimonide was characterized by a bend in the curve at 530°. Preparation for
melting in the solid state was less clearly expressed than in gallium antimonide. Complete severance of the
covalent bonds, judging from thermal effects, ended at ~560°. Heating the melt to 1000° and then cooling until the onset of solidification led to no thermal effects. Apparently the metallic state was retained by the liquid
compound right down to the onset of solidification. Solidification took place at 425−435°, i.e., with supercooling of 95−105°. In the solid state, no thermal effects were observed right down to 20°.

The heats of fusion were determined for each of the substances studied. Our own values and those derived
from calorimetric measurements are given in Table 3.

Analysis of the results obtained shows that in the absence of seeds pure transitional metals may be considerably supercooled. Supercooling of iron by 258° was achieved in [2] by melting 150 g iron under slag, and

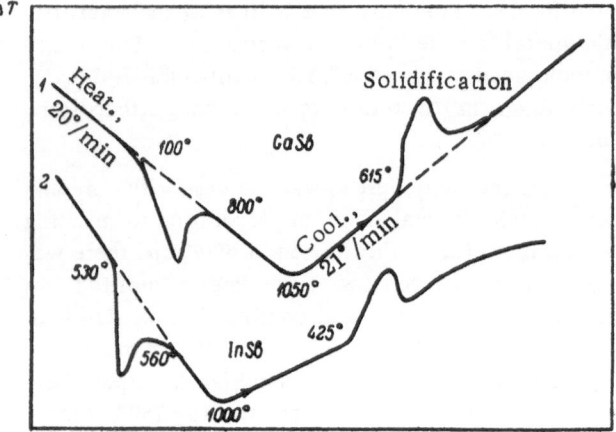

larger degrees of supercooling (350−550°) were
achieved in microvolumes by Dukhin [3], Dukhin
and Niemark [4, 5], and Kamenetskaya, Rakhmanova,
Spektor, and Shiryaev [6]; these findings are supported
by our own data obtained in different experimental
conditions with samples of 0.5−0.3 g. Upon severe
supercooling of molten iron, owing to the small hysteresis
of its polymorphic transformations, solidification may
pass by the δ-phase and take place directly in the γ-
phase. Cobalt and nickel can also be considerably
supercooled, at least by 120−200°. The absence of
supercooling in manganese was evidently due solely
to the lack of specimens of sufficient purity.

The polymorphic transformations of iron, in contrast to solidification, show no great hysteresis on cooling. The maximum supercooling of the centered cubic

Fig. 5. Thermal effects in the phase transformations
of gallium antimonide and indium antimonide.

Element	Melting point, °C	Heat of fusion from author's experimental data, cal/g	Heat of fusion from published data, cal/g	Heating rate, deg/min
Cu	1083	47.82	49.0	42
Ag	960	23.78	24.69	42
Au	1063	16.75	15.46	42
Fe	1540	66.0—66.3	66.02	21—42
In	156	6.8	6.79	42
Sn	232	14.78	13.96	42
Sb	630	43.2	39.01	42
Ge	937	97.128	100.5	21
GaSb	706	86.3		24
InSb	530	44.7	47.2	22

δ−phase reaches 30−40°, and that of the dense-packed cubic γ−phase 40−50°. In iron−manganese and iron−nickel alloys, however, the hysteresis of the γ → α−transformation attains 400−500°C [7]. The fall in the transformation temperatures of body-centered cubic δ−manganese and face-centered cubic γ−manganese is slight. Far more severely supercooled is the composite cubic structure of β−manganese (by 100−200°). The transitional metals show no marked thermal effects which would indicate preparation of the structure for melting ("premelting"), and there are no special signs of a change of short-range order in the liquid state.

On passing over to covalent−metallic elements (gallium, indium, antimony, and bismuth), and especially to semiconducting substances with directional interatomic bonds in the crystalline state (silicon, germanium, gallium antimonide, indium antimonide), the preparation of the crystal for melting leads to an increased absorption of heat before melting begins (premelting), and after melting, in the liquid state, there are thermal effects which indicate a process of gradual disruption of the directional bonds partly retained after transformation into the liquid state. When such substances are further heated, the directional bonds are completely broken, and the substance passes entirely into the metallic state. The latter proves extremely stable, as indicated by the absence of any thermal effects on cooling gallium, indium, silicon, germanium, antimony, bismuth, gallium antimonide, and indium antimonide right down to their solidification temperatures. Supercooling of these substances is slight, and their solidification is accompanied by transformation to the covalent structure, i.e., it coincides with polymorphic transformations, as indicated by the absence of thermal effects after solidification.

Conclusions

1. The polymorphic transformations of face-centered and body-centered cubic structures of manganese and iron on heating take place with insignificant hysteresis and upon cooling, the hysteresis does not exceed 30-50°. The hysteresis of the transformations of composite cubic structures (β → α−transformation of manganese) may reach 200°.

2. The solidification of pure iron, cobalt, and nickel may occur with considerable supercooling; in the present investigation this reached 300, 320, and 130°, respectively. As a result of severe supercooling, iron may skip the δ−phase and solidify directly in the γ−phase.

3. The polymorphic transformations on heating and cooling are marked by clear bends in the curves, while the magnetic transformations in iron, cobalt, and nickel take place gradually, the magnetic transformation temperature constituting the maximum.

4. The melting of certain semiconducting substances having high heats of fusion (silicon, germanium, antimony, GaSb, and InSb) is preceded by preparation for melting ("premelting"), shown by a gradual rise in the curves before reaching the melting point. This is less marked in metals.

5. After the semiconducting substances have melted, they still show thermal effects, apparently due to continuing disruption of directional bonds. This effect is not so noticeable in metals where there is no change in coordination number on melting.

6. On cooling liquid semiconductors which have been heated to high temperatures and have passed into the metallic state, right down to the freezing point, there are no thermal effects which might indicate the appearance of covalent bonds before transformation into the solid state. Solidification takes place with slight hysteresis and leads to the formation of a crystal structure characteristic of semiconductors in the solid state.

7. The values obtained for the heats of fusion of the substances studied — metals and semiconductors — agree with published data. The heats of fusion of gallium and indium antimonides equal 86.3 and 44.7 cal/g.

Literature Cited

1. N. A. Nedumov, High-Temperature Method of Noncontact Thermography, Zh. Fiz. Khim., 34(1):184 (1960).
2. P. Bardenheuer and B. Bleckmann, Mitt. Kais. Wilh. Inst. Eisenforsch., 21:201 (1939).
3. A. I. Dukhin, Solidification of Metals and Alloys in Small Volumes. Problems of Physical Metallurgy and the Physics of Metals. Transactions of the Institute of Physical Metallurgy and the Physics of Metals, No. 6 (TsNIICherMet, Metallurgizdat, Moscow, 1959), p. 9.
4. A. I. Dukhin and V. E. Neimark, Effect of Boron and Titanium on the Supercooling of Steel Ibid., p. 34.
5. V. E. Neimark and A. I. Dukhin, Effect of Modifiers on the Structure of a Steel Ingot, Ibid., p. 39.
6. D. S. Kamenetskaya, É. P. Rakhmanova, E. Z. Spektor, and V. I. Shiryaev, Mechanism of the Effect of Aluminum on the Development of Crystallization Nuclei in Molten Iron, Ibid., p. 63
7. N. N. Sirota, Hysteresis of Phase Transformation, Transactions of the Technological Department of Ferrous Metallurgy, Vol. XIII, p. 108 (1958).

VARIATION OF THE SEGREGATION COEFFICIENT OF IMPURITIES IN GERMANIUM WITH CONCENTRATION

V. I. Korol'kov and V. N. Romanenko

In the technology of preparing semiconductor systems, it is important to have materials of high purity as well as prescribed electrophysical properties—not infrequently those with an assigned volume distribution of an impurity element. The experimental technique of preparing such crystals is based on the passage of the impurity component through the boundary of two phases, liquid and solid, the distribution of the impurity between these being characterized by the segregation coefficient.

In choosing the technological conditions for growing the crystal and in assigning some particular growth program, one must consider the way in which the impurity segregation coefficient depends on the composition of the solid or liquid phase. This is because, in the generally accepted procedures for growing semiconducting crystals, it is always the composition of the phase which is given rather than the temperature for the onset of crystallization. The preparation of tunnel diodes, which have found wide application in radioelectronics, demands crystals with large concentrations of the impurity element, approaching the limit of solubility [1]. In preparing such crystals, and also in producing p—n junctions by the direct fusion method, we need a precise knowledge of the concentration-dependence of the segregation coefficient.

Although phase diagrams for a number of binary alloys of the germanium-impurity type have recently been studied in detail [2, 3] both experimentally and theoretically, nevertheless up to the present the direct concentration-dependence of the segregation coefficient of impurities in semiconductors, of great practical importance, has not been examined.

It should be noted that in studying phase diagrams of the semiconductor—impurity type it is the equilibrium phase diagrams which are of chief interest. Under the actual conditions of growing a crystal from an impurity-doped melt, one has to deal with a nonequilibrium phase diagram. We know from experimental investigations of small impurity concentrations that there may be a considerable difference between equilibrium and actual segregation coefficients [4]. For practical purposes we must therefore study the concentration-dependence of the segregation coefficient under the conditions found in actual practice.

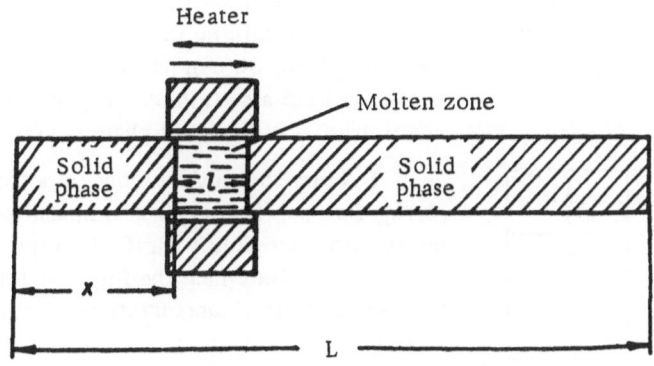

Fig. 1. Schematic representation of the zone equalization
process.

We investigated the concentration-dependence of the segregation coefficients of gallium and antimony in germanium. The surface of phase separation was moved at a rate of ~1.5 mm/min.

In determining the value of the segregation coefficient we used the method of zone equalization, enabling us to obtain a uniform distribution of impurity along the length of the bar which introduced the impurity into the leading zone. In this zone equalization process, as in ordinary zone melting, a molten zone is formed in a long sample (Fig. 1).

In contrast to the zone refining condition, however, this zone moves through the samples alternately backward and forward. We know that for a fair number of such to-and-fro passages the zone becomes "impurity saturated," and that on its further motion the composition of the liquid phase, and hence also that of the solid phase forming at the solidification front, remains constant. Thus a certain limiting distribution of impurities along the length of the sample is reached. It may be shown that the limiting distribution is constant if there are no regular changes in the length of the zone as it moves [5]. Thus all the impurities presented in the sample are distributed quite evenly over its length, except in the place at which the liquid zone lies. The impurity concentration here is connected with the concentration elsewhere by the segregation coefficient. After the zone equalization process is terminated, the zone passes slowly out of the sample at one end.

Here the zone length is curtailed, and the impurities in this section fall out on the normal cooling law [6]. The limiting concentration in the solid phase may be expressed, from the law of conservation of total impurity, in terms of the zone length and the total impurity in the sample [7]:

$$C_\infty = \frac{\int_0^A kC_0(a)\,da}{1 + k(A - 1)}, \tag{1}$$

where k is the segregation coefficient, A and a are dimensionless coordinates, $A = L/l$, $a = x/l$, L = sample length, l = zone length, and $C_0(a)$ is the initial distribution of impurity concentration along the length of the sample. The x axis is along the sample, and the origin of coordinates is at one end. The position of the zone is characterized by the solidification point.

It was shown in [7] that the limiting distribution of impurity is reached very rapidly in most cases. Two or three to-and-fro passages are sufficient for the resultant distribution to differ very little from the limiting state.

This was also confirmed experimentally [8]. Knowing the size of the molten zone, the length of the sample, and the total amount of impurity introduced, and determining the limiting impurity distribution in the solid phase, we can not only calculate the segregation coefficient but also immediately determine to what impurity concentration in the solid phase it corresponds.

In order to study the concentration-dependence of the segregation coefficient, it is simplest to introduce a known impurity into a previously purified sample. It is technologically convenient to introduce the impurity into the leading zone. The form of the initial concentration of impurity is in this case written as

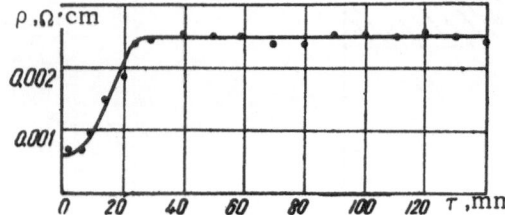

Fig. 2. Distribution of specific resistance along the bar after three cycles of zone equalization.

$$C_0(a) = \begin{cases} C_i & 0 \leqslant a \leqslant 1 \\ 0 & 1 < a \leqslant A. \end{cases} \tag{2}$$

The limiting distribution is given by the formula

$$C_\infty = C_i \frac{k}{1 + k(A - 1)} .\tag{3}$$

Hence it follows that the segregation coefficient is

$$k = \frac{1}{\dfrac{C_i}{C_\infty} - A + 1} .\tag{4}$$

Experimental Part

In order to study the concentration-dependence of the segregation coefficient of impurities in germanium we chose Ge which had been previously purified to a resistivity of 50−60 Ω·cm, i.e., possessing intrinsic conductivity at room temperature. This Ge purity was needed in order to offset the influence of other impurities which might distort the picture of the actual impurity distribution.

Figure 2 shows the distribution of specific resistance after three cycles of zone equalization. The four-probe method was used to measure the distribution of specific resistance along the bar so as to reveal the part (equal to the zone length) crystallizing on the normal cooling law. The part at which the zone passed out was cut off and the volume of the liquid zone V_l and the remaining part of the bar V_s were determined in order to give the value of A, allowing for the change in the density of Ge on melting [9]. The values of the remaining quantities in Eq. (4) were easy to determine. The value of C_i could be calculated, knowing the weight of impurity introduced and the volume of the initial zone.

The concentration of impurity in the solid phase, equal to the limiting concentration C_∞, was determined by means of Hall measurements. Since at room temperature all the donors or acceptors (elements of the fifth and the third group of the periodic table) are completely ionized, to each impurity atom there corresponds one carrier and the impurity concentration in the solid phase equals the carrier concentration.

As impurities for study we selected gallium Ga−O, and antimony Sb−extra. All the melts with gallium were carried out in vacuo. The melts with antimony were carried out in a sealed ampoule filled with spectroscopically pure argon.

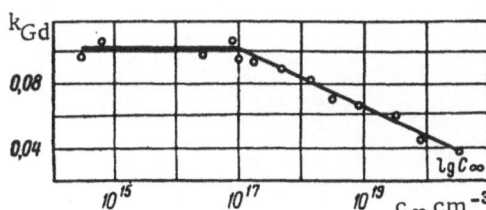

Fig. 3. Concentration-dependence of the segregation coefficient of gallium in germanium.

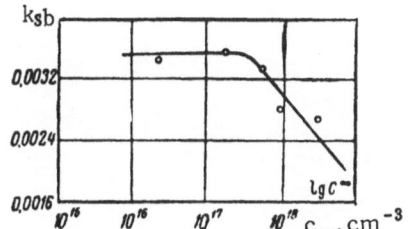

Fig. 4. Variation of the segregation coefficient k_{Sb} of antimony in germanium with the concentration of antimony in the solid phase.

Discussion of Results

We studied the variation of the segregation coefficient k_{Sb} of antimony and k_{Ga} of gallium in germanium with impurity concentration over a wide concentration range. The results are shown in Figs. 3 and 4.

We see from the figures that the values of k_{Sb} and k_{Ga} in the low-concentration region agree closely with the commonly accepted segregation coefficients [10]. Starting from a concentration of the order of 10^{17} atoms/cm^3 for Ga (and somewhat later for Sb), however, there is a fall in the segregation coefficient proportional to $\lg C_{\infty}$. This is evidently connected with a change in the penetration energy of the impurity element into the lattice of the solvent crystal. The fact that in this concentration range, owing to the interaction of impurity atoms, the wave functions of the electrons begin to overlap and form an impurity band supports the idea that impurity atom interaction is responsible for the fall in the segregation coefficients.

It should be noted that the method indicated here for studying the concentration-dependence of the segregation coefficient is simple and reliable. The precision of the segregation coefficient values is determined by that of the electrical measurements. The present method may be successfully used for studying the concentration-dependence of the segregation coefficients of other impurities.

Literature Cited

1. L. Esaki, Phys. Rev., 109(2):603 (1958).
2. F. A. Trumbore, Bell Syst. Tech. J., XXXIX:205, (1960).
3. K. F. Lehovee, Phys. Chem. Solids, 23:695 (1962).
4. R. N. F. Hall, Phys. Chem., 57:836 (1953).
5. B. A. Volchok and V. Ya. Frenkel', Fiz. Tverd. Tela, III:2011 (1961).
6. W. G. Pfann, Trans. AIME, 194:747 (1952).
7. V. N. Romanenko, Fiz. Tverd. Tela, I:1679 (1959).
8. V. N. Romanenko and G. V. Nikitina, Collection: Crystallization and Phase Transformations (Izd. Akad. Nauk Belorus. SSR, Minsk 1962).
9. W. G. F. Pfann, J. Metals, 5(11):1441 (1953).
10. J. A. Burton, Physica, XX:845 (1954).

STUDY OF HETEROGENEOUS EQUILIBRIUM
DURING THE CRYSTALLIZATION OF GERMANIUM
FROM MELTS CONTAINING ELEMENTS
OF THE DONOR AND ACCEPTOR TYPES

A. D. Belaya and V. S. Zemskov

The study of the equilibrium of liquid and solid phases is based upon the method of comparing the concentrations of elements in the primary crystals of a solid solution with their concentrations in the liquid phases from which the crystals were obtained. In order to obtain primary germanium crystals from melts containing various quantities of doping elements, we used the method of drawing the solid phase from the melt [1,2].

The drawing was effected at a rate of 0.04 mm/min, the crystal being rotated at 170 rpm and thus intensively mixing the melt. Good mixing of the melt and a slow drawing rate ensured nearly equilibrium conditions for crystallization of the solid phase. The germanium used in the experiments had a specific resistance of 30 $\Omega \cdot$cm, an electron mobility of 3600 cm^2/V$\cdot$sec, and a diffusion length of the minority carriers of not less than 2 mm. The doping elements—aluminum, antimony, indium, and gallium — contained less than $10^{-4}\%$ impurity. The drawing was carried out from melts with compositions lying on the sections of constant germanium composition. The sections corresponded to the region of primary precipitation of germanium-base solid solution crystals.

By measuring the Hall constant on the samples obtained, we were able to determine the total concentration of doping elements. The concentration of one of the doping components was determined by quantitative radiography, which eliminated the introduction of errors by the inclusion of a doping component not entering solid solution. The radioisotopes Sb^{124} and In^{114} were used for this purpose. The amount of the second doping component was calculated from the difference between the total concentration calculated from the Hall constant measurements and the concentration of the "tagged" component given by the radiographical measure; ments. Three samples were prepared and studied for each concentration of doping components in the melt. All the samples were single crystals and were single-phase under microscopical examination. The methods of investigation are described in detail in [1, 2]. Among systems studied were germanium—aluminum—antimony, in which the doping elements were of different types, donor and acceptor, and germanium—indium—gallium, in which both indium and gallium were acceptors.

The Germanium — Aluminum — Antimony System

In studying this system, we drew samples of solid solutions from melts belonging to three sections: 97, 90, and 80 at. % Ge. The compositions of the original germanium melts, with an indication of the amount of aluminum and antimony in them and the concentration of the doping components in the resultant solid phase, are shown in Figs. 1–3. The concentrations of aluminum and antimony in the solid phase are shown on different scales in the figures.

We see from the figures that the compositions of the solid phases solidifying out of melts belonging to the germanium—aluminum antimonide section do not lie on the same section. In equilibrium with the melts corresponding in composition to the germanium—aluminum antimonide section are phases in which there is more aluminum than antimony. In order to have an equal amount of aluminum and antimony in the solid phase, i.e., an equiatomic relationship between aluminum and antimony, the melts must be made richer in antimony than in aluminum.

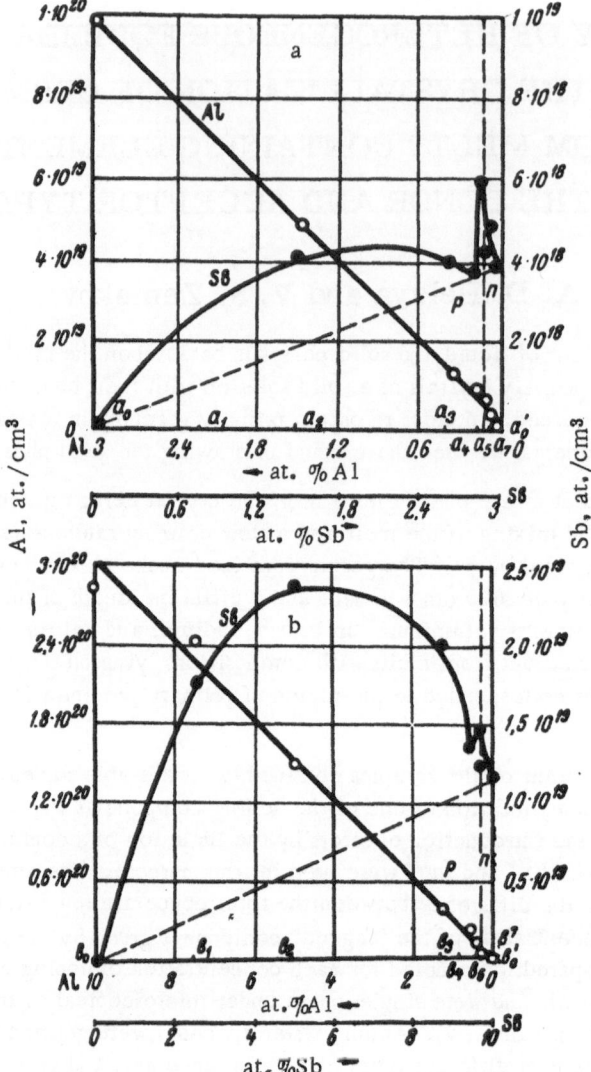

Fig. 1. Variation of the concentration of aluminum and antimony in the solid phase with their concentration in the melt: a) For section 97 at. % Ge (a−a); b) 90 at. %, ● antimony; ○ aluminum.

Thus, as in [2], the above results convincingly indicate that the germanium−aluminum antimonide section does not reflect the equilibrium existing between liquid and solid phases on crystallization of a melt having a composition lying on the germanium−aluminum antimonide section. The relation between the amounts of aluminum and antimony in the melt needed to ensure an equiatomic relation in the solid on crystallization may be predicted if we know the distribution coefficients of these elements in germanium, taken for corresponding temperatures and a single doping element.

In the study of the analogous systems germanium−indium−antimony [1] and germanium−aluminum−arsenic [3], it was also observed that the germanium−$A^{III}B^{V}$ sections did not reflect equilibria existing between the liquid and solid phases for the crystallization of melts lying in these sections.

Figures 1-3 also show the straight lines along which the aluminum and antimony concentrations would have to move if interaction between the doping elements were completely absent, i.e., if the aluminum and antimony entered into the germanium in accordance with the distribution coefficients proper to these elements in the case of separate doping.

Despite the fact that the selections studied were not isothermal, the construction of such hypothetical straight lines was justified, since analysis of the liquidus lines of the germanium—aluminum, germanium—antimony, and germanium—aluminum antimonide systems showed that the difference in the liquidus temperatures for these systems, within the limits of the concentrations considered, were insignificant, and could not markedly affect the variation of the distribution coefficient. Hence the assumption that the distribution coefficients of the doping components remain constant for all alloys of each section introduces no serious error.

Comparing these hypothetical lines with those found by experiment, we can see that the concentrations obtained are somewhat larger than the hypothetical. The increase in concentration is more obvious for antimony than for aluminum. Moreover, the change is the greater, the lower the solidification temperature of the solid phase. The observed change in concentration indicates the presence of interaction between the atoms of the doping components. The most probable cause of the change in the concentration of aluminum and antimony in the solid phase is the change in the conditions of ionization of the atoms of these elements in the solid phase upon the altering of the ratio of the doping components [4]. In this case the distribution coefficient of the doping component for twofold doping may be expressed [5] as

$$K = K^* \left(\frac{\gamma_l}{\gamma_l^*} \right) \left(\frac{Fn^*}{Fn} \right),$$

(1)

where K^* is the distribution coefficient of the doping component in germanium on doping with only one element, γ_l, γ_l^* are the activity coefficients of the doping component in the liquid phase in ternary and binary solutions, respectively, and Fn^*, Fn are the proportions of atoms of the doping component existing in the neutral state in binary and ternary solid solutions, respectively.

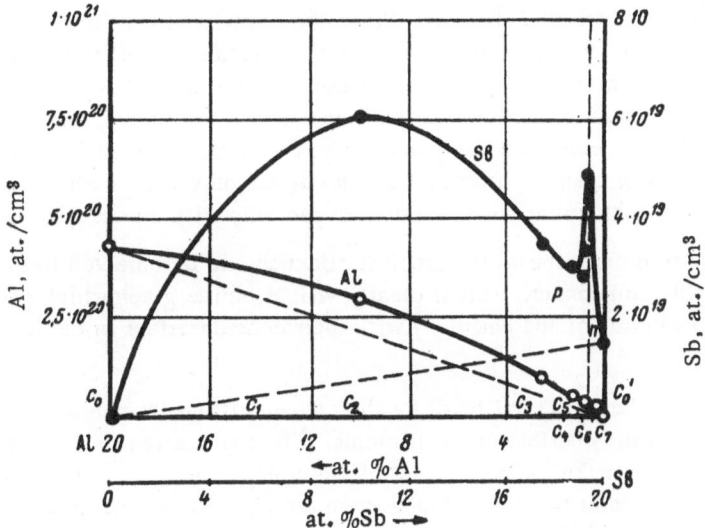

Fig. 2. Variation in the concentration of aluminum and antimony in the solid phase with their concentration in the melt for the section 80 at. % Ge (c—c): ● antimony; ○ aluminum.

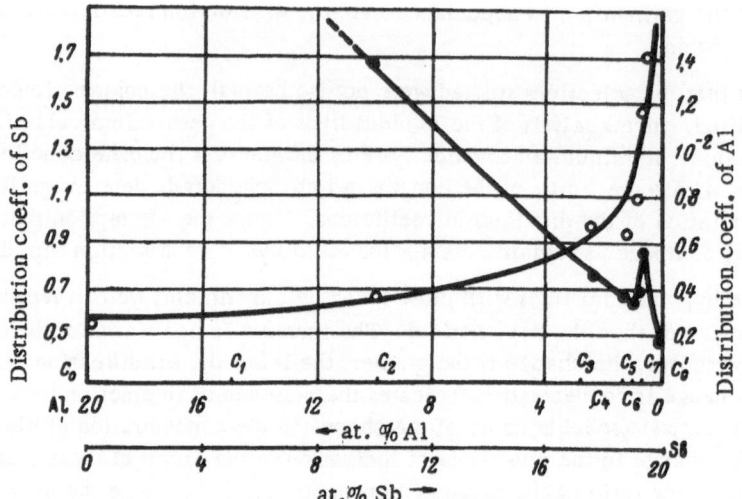

Fig. 3. Variation of the distribution coefficients of aluminum and antimony with their concentration in the liquid phase for the section 80 at. % germanium: ● antimony; ○ aluminum.

These proportions can be calculated if the Fermi level is known [6]. For the acceptor,

$$Fn = \frac{1}{1 + [0.5 \exp \{(E_F - E_A)/KT\}]} ,$$

(2)

where E_F is the Fermi level and E_A is the ionization energy of the acceptor.

An analogous relation holds for donors, with the difference that the energy term takes the form $(E_D - E_F)$, where E_D is the ionization energy of the donor. Assuming that in Eq. (1) the ratio $\gamma_L/\gamma^{\bullet}_L \cong 1$, we may estimate the change in the distribution coefficient. Since the position of the Fermi level is governed by the number of dissolved atoms of the doping components, and the separate solubilities of aluminum and antimony in the solid phase for the temperatures in question differ considerably from one another, then, on varying the ratio of aluminum and antimony atoms in the solution (and hence in the solid phase) from that of a solution containing only aluminum to one containing only antimony, the Fermi level shifts from the position corresponding to the concentration of donor element atoms in samples doped only with a donor to that corresponding to the concentration of acceptor element atoms in samples doped only with an acceptor.

For the given interaction mechanism, the greatest effect should be observed in the case of maximum dilution of the second doping component. This is clearly visible on the graph which gives the variation of the distribution coefficients of aluminum and antimony with their concentrations in the liquid phase for the 80 at. % germanium section (Fig. 3).

Using Blakemore's [7] values of Fermi level for the concentrations investigated, we may estimate the maximum expected change in the distribution coefficients. The calculation shows that, on diluting the aluminum with antimony, the distribution coefficient of aluminum should rise, and for an insignificant amount of aluminum in the melt it becomes twice that found when the germanium is doped with aluminum alone, for the 97 at. % Ge section, roughly twice for the 90 at. % Ge section, and three times for the 80 at. % Ge section.

The distribution coefficient of antimony on diluting with aluminum also rises by factors of 4, 8, and 15 compared with that obtained when the germanium is doped with antimony alone for the 97, 90, and 80 at. %

Ge sections, respectively. The calculated data are in good agreement with experiment. The coincidence between experiment and theory confirms the view that a change in the ionization conditions of the atoms of the doping components is a leading factor in changing the concentration of these doping elements.

We see from Fig. 1—3 that, apart from a general tendency to increase the concentrations of antimony and aluminum in the region where the elements are in equiatomic ratio in the solid phase, a certain maximum also occurs, increasing with falling temperature. This maximum cannot be explained by the mechanism described. The increase in the solubility of these elements is probably caused by the formation of a complex [Al Sb] in the solid phase. There is no basis for the supposition that these complexes can form in the liquid phase, since in this case the antimony atoms in the liquid phase are much larger (~25 times) than the aluminum atoms. When the aluminum and antimony are in equiatomic proportions in the liquid phase, there is no such maximum. This explanation, however, requires checking. Such an increase in the concentration of doping elements when contained equiatomically in the solid phase was also observed in [8], where the method of quenching the alloys was

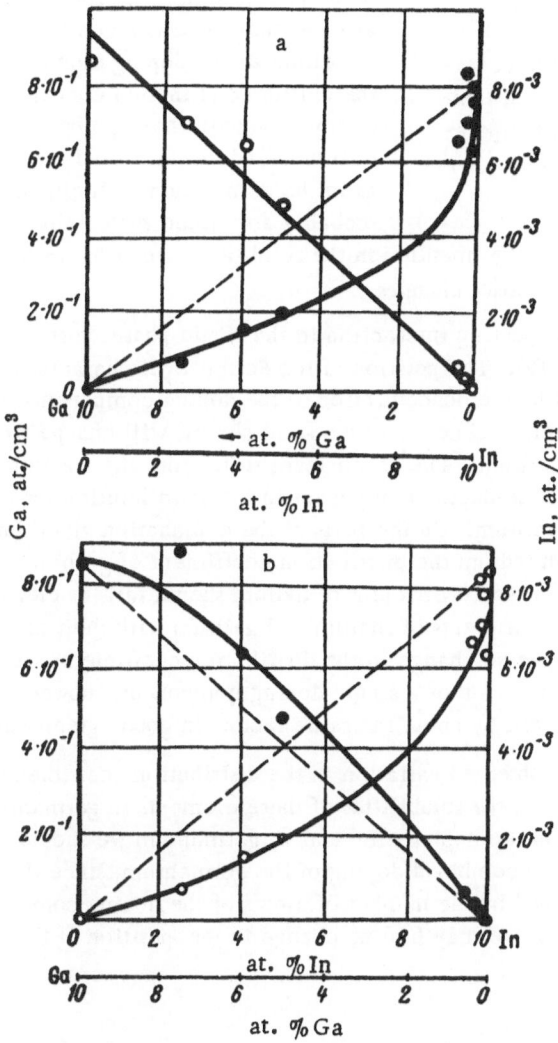

Fig. 4. Variation in the concentration of gallium and indium in the solid phase with their concentration in the melt at temperature 909°C: ● gallium; O indium.

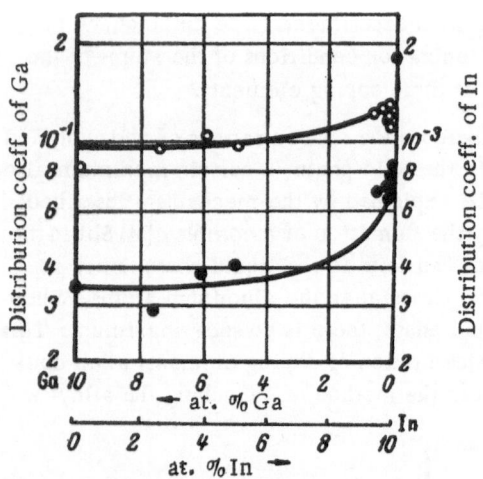

Fig. 5. Variation in the distribution coefficient of gallium and indium with their concentration in the melt at a temperature of 909°C: ● gallium; ○ indium.

used in studying the germanium−aluminum−antimony system.

Germanium − Indium − Gallium System

In the germanium−indium−gallium system we studied samples of solid solutions drawn from melts belonging to the section with 90 at.% germanium. Thermal analysis of the alloys studied showed that this section was isothermal. The temperature of the beginning of solidification of the melts was 909°C. The compositions of the original germanium melts, with an indication of the quantity of indium and gallium atoms in them and the concentrations of the doping components in the resultant solid phases, are given in Fig. 4. The figure also contains hypothetical straight lines for the variation in solubility when one assumes that there is a complete absence of interaction between the atoms of the doping elements. We see from the graphs that the gallium and indium concentrations behave differently in comparison with the hypothetical straight lines. The concentration of indium falls, and that of gallium tends to rise; just as in the germanium−aluminum−antimony system, this is probably associated with a change in the ionization conditions of the indium and gallium atoms in germanium in the solid phase when the ratio of these atoms present in the melt (and hence in the solid phase also) changes.

There are no grounds for expecting interaction in the liquid phase; germanium−indium and germanium−gallium melts are close to ideal [9]. The position of the Fermi level characterized by the ionization conditions of the atoms will be determined by the concentration of the doping components, which, on varying of the ratio of the doping elements in the melt (and hence in the solid phase), will change from a melt doped only with indium to one doped only with gallium. The Fermi level shifts from the position corresponding to a concentration of $3.8 \cdot 10^{18}$ atoms/cm^3 of the doping component in the solid solution for indium to a position corresponding to $3.8 \cdot 10^{20}$ atoms/cm^3 for gallium. On the basis of the explanation given for the change in concentration, upon dilution of the gallium with indium the distribution coefficient of gallium should rise, and upon dilution of indium with gallium the distribution coefficient of indium should fall, as clearly seen in the graphs giving the variation in the distribution coefficients of indium and gallium with their atomic ratios in the melt (Fig. 5). Calculation of the maximum expected change in the distribution coefficients, as compared with those of the same elements in germanium in the case of a single doping component, showed that that of gallium rises by a factor of ∼2 and that of indium falls by about the same factor, in good agreement with experiment.

It should be noted that the observed variation in the distribution coefficients of elements of one type is explained by the large difference in the solubilities of these elements in germanium. If there were no difference between the solubilities of the doping elements in germanium we should not expect a change in the concentration of these elements on combined doping of the germanium, since the position of the Fermi level producing this change is determined by the number of atoms of the doping components. For solubilities either the same or differing very slightly, we may find no change in the position of the Fermi level, as indeed was observed in [10].

Conclusions

1. It is quantitatively confirmed that the germanium−aluminum antimonide section in the germanium−aluminum−antimony system does not reflect equilibrium between the liquid and solid phases at the temperatures studied.

2. .The mutual influence of the doping additives on their distribution in germanium in the germanium—aluminum—antimony and germanium—indium—gallium systems is established. Upon the simultaneous doping of germanium with aluminum and antimony, the distribution coefficient of aluminum in germanium rises by factors of 2, 2, and 3, and that of antimony by factors of 4, 8, and 15 for the 97, 90, and 80 at.% Ge sections, respectively. On simultaneously doping germanium with indium and gallium, the distribution coefficient of gallium in germanium at 909°C rises by a factor of two, and that of indium falls by the same factor as compared with the distribution coefficients of these elements in germanium for the case of single doping.

3. The change in the distributions of the doping components in germanium on composite doping is caused by the variation in the ionization conditions of the atoms of doping components with the position of the Fermi level.

4. A sharp rise in the concentrations of aluminum and antimony in the germanium—aluminum—antimony system when they are equiatomically present in the solid phase must be assumed to be due to the formation of [AlSb] complexes in the solid state.

Literature Cited

1. V. S. Zemskov, B. G. Zhurkin, and A. D. Suchkova, Study of Heterogeneous Equilibrium in the Germanium—Indium—Antimony System, Zh. Fiz. Khim., (9):36 (1962).
2. V. S. Zemskov, A. D. Suchkova, and Van-Gui-Hua, Nature of the Heterogeneous Equilibrium in the Germanium—Aluminum—Antimony System, Izv. Akad. Nauk SSSR, Otd. Tekh. Nauk, Metal. i Toplivo, No. 6 (1961).
3. J. O. McCaldin, Interaction between Arsenic and Aluminum in Germanium, Appl. Phys. 31(1): (1960).
4. H. Reiss, Chemical Effects due to the Ionization of Impurities in Semiconductors, J. Chem. Phys., Vol. 21. (7):1209—17 (1953).
5. C. D. Thurmond, Distribution Coefficient of Impurities Distributed between Ge or Crystals and Ternary Alloys or Surface Oxides. Properties of Elemental and Compound Semiconductors (New York, 1959), pp. 121—40.
6. N. B. Hannay, Semiconductors (New York, 1959), 195—215.
7. J. S. Blakemore, The Fermi Level in Germanium at High Temperatures, Proc. Phys. Soc., 71(460):692 (1958).
8. V. M. Glazov, D. A. Petrov, and S. N. Chizhevskaya, Simultaneous Solubility of Elements of the Third and Fifth Groups in Germanium, Izv. Akad. Nauk SSSR, Otd. Tekh. Nauk, Metal. i Toplivo, No. 4 (1959).
9. C. D. Thurmond and K. M. Kowalchik, Germanium and Silicon Liquidus Curves, Bell Syst. Techn. J., 39:169—204 (1960).
10. F. A. Trumbore, E. M. Porbansky, and A. A. Tartagalia, Solid Solubilities of Aluminum and Gallium in Germanium. Phys. Chem. Solids, 11(3/4) 239—45 (1959).

GROWTH OF GERMANIUM CRYSTALS FROM A MELT
CONTAINING A CONSIDERABLE AMOUNT OF IMPURITY

A. Ya. Gubenko

Certain features of the growth of crystals from a melt containing a considerable concentration of impurity afford a clear picture of crystal growth forms. During growth from such melts, the tangential velocity of a step on the surface extending in the direction of growth will decrease with increasing concentration of impurity in the melt. Steps moving at higher speeds will overtake the slower ones and merge with them, thus increasing the height of the step being formed. The distribution of impurity in the melt will be inhomogeneous over the height of the step. At the incoming corner of the step the concentration of impurity will be greater than at the leading edge, from which the impurity passes off more rapidly into the depths of the melt. The existence of a concentration (and hence temperature) gradient at the step will enhance the height and the stability of the increase. Since the probability of forming a second phase is determined by the concentration of impurity in the melt and the degree of supercooling at the crystallization front, the macrostep being formed will be conducive to the formation of a second phase at the crystallization front.

The second phase being formed delineates the contour of the face which leads the crystallization. Decoration of the principal face takes place. When the decorating second phase layer is etched away, the form of the growing crystal can be easily determined.

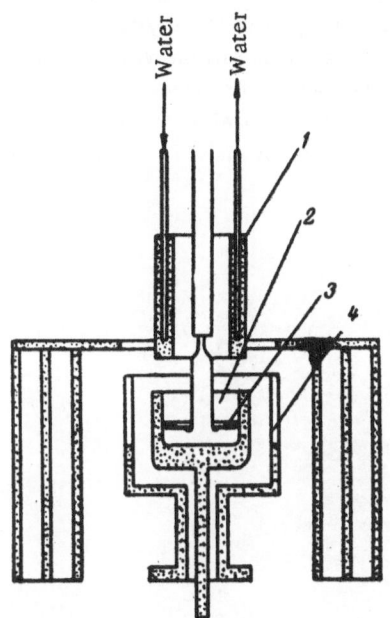

Fig. 1. Schematic representation of a system for growing a germanium single crystal: 1) Cooling system; 2) Crucible; 3) Filter; 4) Heater.

For somewhat smaller impurity concentrations in the melt, when the probability of forming a second phase is slight, if the tangential velocity of solidification varies, again the forming layer will have a concentration considerably different from that of the substrate on which it was formed. Layers with different interatomic spacings will develop, and these will be separated by rows of dislocations. Since the dislocations are disposed along the growing face, we shall also be able to judge the growth form from etch pits.

Thus the development of tangentially propagated macrosteps upon heavy doping together with the formation of conjugate layers with different concentrations offer the possibility of observing the growth form in growing crystals, either by metallographic study, or on the end surface of a crystal taken quickly from the melt during growth. These principles have in fact been used for delineating growth forms.

Germanium crystals were doped with gold, arsenic, and gallium while being grown by Czochralski's method. In order to reduce concentration supercooling and obtain a plane front, the crystals were pulled with a large temperature gradient and a high seed rotation rate. A large temperature gradient was created through forced cooling of the growing crystal by placing a cooling system around it (Fig. 1). In order to increase heat transfer from crystal to cooler, the process was carried out in a helium atmosphere (large thermal conductivity), and to increase the heat flow from the

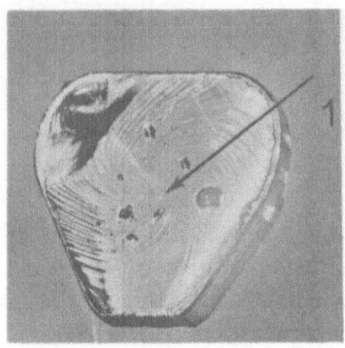

Fig. 2. Crystal surface torn off from the melt. Arrow 1 shows the zone of formation of macro-spirals.

crystal in the growth direction thick seeds of 6 mm diameter were used. The seed rotation rate was $80-100$ rpm, and the crucible rotation rate $3-4$ rpm. The drawing rate was 1.5 mm/min. The temperature gradient at the surface of separation in the solid phase was $150°C$. Crystals were grown in the directions [111], [110], and [100]. When the crystal was broken off from the melt, the drawing rod of the system was cut off from the moving mechanism and, together with the crystal, rapidly pulled upward by hand. The crystal did not pass out of the screen-enclosed field of the furnace. In order for the general background of dislocation etch pits not to "smear" the character of the dislocation pit distribution, due to the inhomogeneous absorption of impurity, crystals with low dislocation densities were grown. In arsenic- and gold-doped crystals grown in the system shown in Fig. 1, the dislocation density was no greater than 10^2 cm^{-2}. Crystals containing different quantities of impurity were cut along a plane perpendicular to the growth axis and etched to reveal the character of the etch pit distribution in a standard etchant, or else etched in KOH and CP-4 for selective etching of the second phase. The breakoff surface of the crystals corresponding to a gold concentration of $5 \cdot 10^{14}$ cm^{-3}, was smooth for growth in the [111] direction. For concentrations of $7-8.10^{14}$cm^{-3}, macrospirals were observed on the broken surface; these reproduced the picture characteristic of the development of screw dislocations of the same kind from a single line source (Fig. 2). In the zone where the development of macrospirals begins, there are elevations constituting their centers of development (Fig. 3). The macrospirals consist of several semicircles or angular steps (light bands). The merging of these steps also leads to the formation of macrospirals. Analogous macrospirals formed by dislocation etch pits were observed in arsenic-doped crystals (Fig. 4). Whereas in a large part of the crystal the dislocation density was insignificant (dark field), at the edge of the crystal, where a change in the velocity of crystallization led to the development of layers with different impurity concentrations, macrospirals were observed. In the center of the crystal, on etching in CP-4, centers of spiral development similar to those shown in Fig. 3 were observed. In Fig. 4, apart from circular macrospirals, we also have straight ones, i.e., the

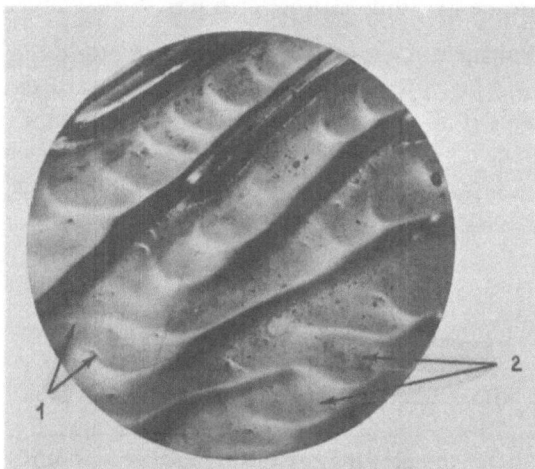

Fig. 3. Surface of crystal torn off from the melt. Arrow 1 shows center of formation of macrospirals, arrow 2 a coagulation of steps with the formation of a single macrospiral. Magnification × 300.

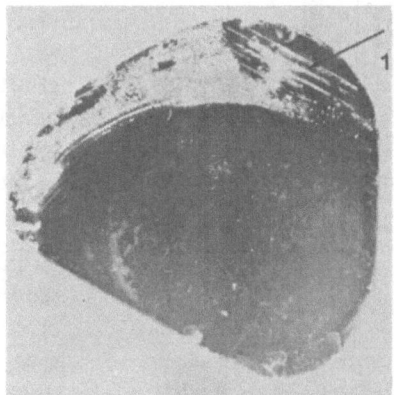

Fig. 4. Crystal surface etched for dislocation density in standard etchant. Magnification × 7. 1) Rectilinear macrospiral.

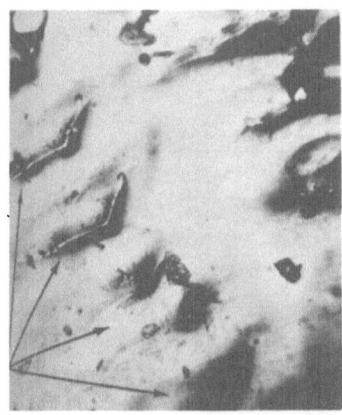

Fig. 5. Surface of germanium crystal etched in CP-4 etchant. Magnification × 300.

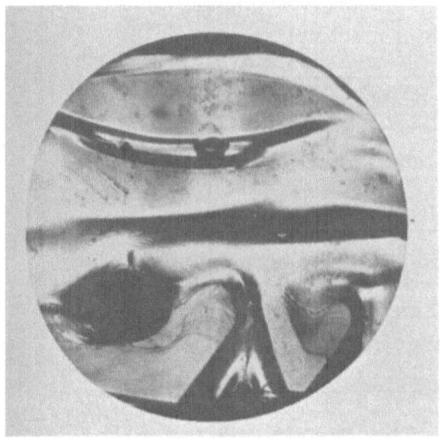

Fig. 6. Formation of octahedra from a new growth center on a macrospiral. Magnification × 300.

circular macrospirals degenerate into straight lines. Upon etching of the same crystal in CP-4, the straight spirals were found to consist of angular steps, the height of which increased as a result of a local increase in impurity concentration (Fig. 5).

These steps are analogous to those composing the macrospirals shown in Fig. 3. On further increase of the impurity concentration in the melt, the macrospirals assumed angular shape, and then new growth forms developed around them (Fig. 6). The latter developed from new growth centers and had the equilibrium germanium form — that of octahedra. For still larger concentrations, the entire surface of the end of the crystal consisted of hexahedra (truncated octahedra), forming a cellular structure. On the outside surface of the crystals in the region of the edge, a series of parallel projections appeared. For growth in the [111] direction, these projections made an angle of 70.5° with the section surface, and for growth in the [110] and [100] directions they were parallel to the direction of growth.

The projections are macrotraces of the [111] planes emerging at the edges of the crystals. For small concentrations, in the upper part of the crystal, the length and height of the projections were not great. Upon increase of the impurity concentration along the length of the growing crystal, the height and length of the projections increased (Fig. 7). For crystals grown in the [111] direction, the projections appeared at higher concentrations than on crystals grown in the [110] and [100] directions. Thus the concentration of arsenic in the crystals corresponding to the appearance of projections for growth in the [111] direction was $2.4 \cdot 10^{19}$ cm^{-3} and for growth in the [100] direction $4 \cdot 10^{18}$ cm^{-3}. Maximum concentrations of impurities were also obtained for growth in the [111] direction. The projections on crystals grown in the [111] direction were extended on the breakoff surfaces by planes bounding the octahedra. Analogous pictures were observed for growth in other directions. All the forms revealed were bounded by [111] planes. The form of dividing boundary and its degree of curvature can determine the form of the growth figures, but not change their nature. Thus in the case of a convex growth front the dislocation pits formed macrospirals in the form of closed loops.

The table presents some data indicating the variations in growth form as the impurity concentration in the crystals changes. For impurity concentrations differing by an order from the limiting values, the breakoff surfaces were smooth, there were no projections, and the dislocation pits did not form characteristic series. This is probably due to the small height of the step and the insignificant effect of impurity on the tangential growth velocity. With increasing impurity concentration macrospirals appeared, and when the concentration supercooling reached a substantial value new growth centers were formed on the spiral and a cellular structure developed. With reduction of the temperature gradient (removal of the cooling system) all the concentrations for the respective growth forms tended to decrease.

Thus, when germanium crystals are grown with a considerable impurity concentration from the melt, macrospirals are formed; these originate in the coagulation of layers moving at different tangential velocities.

Dependence of the Form of Growth of the Crystals on the Character on the
Character and Concentration of Impurities

Doping impurity	Conc. in solid phase (cm^{-3})	Direction of growth	Observed forms of growth
Au	$5 \cdot 10^{14}$	[111]	Breakoff surface smooth. No grooves.
Au	$6 - 7 \cdot 10^{14}$	[111]	Macrospirals. Projections appeared
Au	$7 - 8 \cdot 10^{14}$	[111]	Macrospirals. Projections. Individual octahedra
Au	$> 8 \cdot 10^{14}$	[111]	Cellular structure. Large depth projections
As	$4 \cdot 10^{18}$	[110]	Projections. Macrospirals
As	$4 \cdot 10^{18}$	[111]	No projections or macrospirals observed
As	$3 \cdot 10^{19}$	[111]	Projections appeared.
As	$3 \cdot 10^{19}$	[110]	Cellular structure. Projections
As	$4 - 5 \cdot 10^{19}$	[111]	Projections, macrospirals
As	$-6 \cdot 10^{19}$	[111]	Angular macrospirals. Projections
As	$> 6 \cdot 10^{19}$	[111]	Cellular structure. Large depth projections

As the concentration rises, the macrospirals lose their round form and become angular. The transition to the angular form, the formation of projections on the outer surface of the crystals, the transition to octahedral forms, and the formation of a cellular structure are due to the creation of conditions favorable to the growth of equilibrium forms of germanium. These conditions arise, according to Ellis [1], when the temperature gradient in the

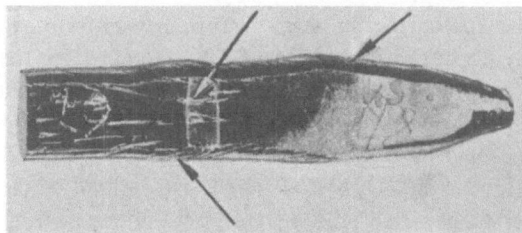

Fig. 7. Gallium-doped crystal. Arrows show projections. In the lower
part the height and length of the projections are enlarged.

melt and the distribution of equilibrium crystallization temperatures differ little from one another. A transition from one growth form to another is determined by the concentration of the doping impurity at the dividing boundary, the temperature gradient, and the crystallographic direction of growth. Since all the observed growth figures, independent of growth direction, are bounded by [111] planes, the angles between the [111] plane and the growth front and the form of the steps moving tangentially will affect the growth, as well as the crystallography of the growing surface.

A maximum concentration of impurity can be obtained by growing crystals in the [111] direction. Since the form of the growth figures in some measure determines the impurity distribution, the choice of growth conditions determines the inhomogeneity of the electrophysical properties of the resulting material.

Literature Cited

1. J. Ellis, J. Appl. Phys. 26(9):440 (1955).

METHOD OF DECORATING THE STRUCTURE OF SILICON BARS
BY FUSING A VERY THIN SURFACE LAYER

N. Shamba

When an extremely thin surface layer of a silicon bar is fused, a macroscopic "rash" appears on the surface; this is formed by external fine blisters arranged in a definite order, which depends on the crystallinity of the bar.

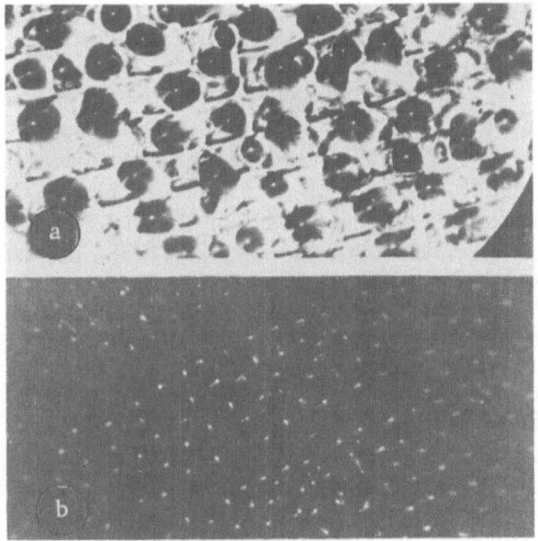

Fig. 1. Network of blisters in a silicon single crystal at different magnifications: a) × 2000; b) × 80.

The Czochralski crystal-drawing system creates conditions reducing the temperature gradient between the region above the melt surface and the "overhead" position of the bar in the apparatus.

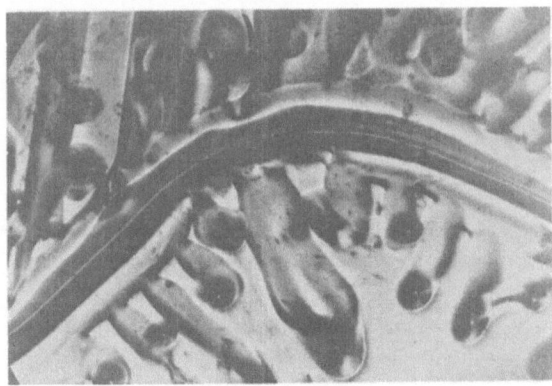

Fig. 2. Blisters on a silicon dendrite.

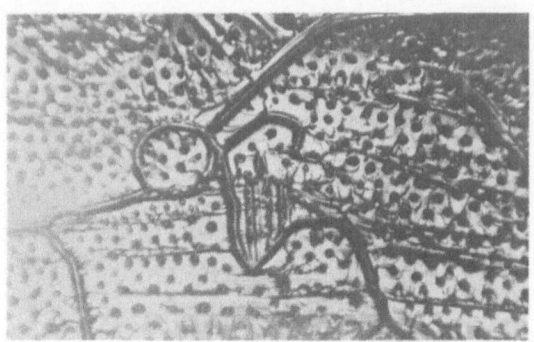

Fig. 3. Network of blisters on a polycrystalline
silicon bar.

The bar being drawn is set in the "overhead" position and held there for a period of time to establish temperature equilibrium at 1200°C. Then with a rapid movement it is let down (by hand) into the region above the surface of the superheated melt, and instantly raised back to the "overhead" position. As a result, there appears on the surface of the bar a bright fused film covered with a "rash" of blisters.

In the case of a single-crystal bar, this film is covered very regularly with a network of blisters (Fig. 1a and b). Here the blisters are fairly equidistant and all are of roughly the same size.

Upon the fusing of a dendrite (Fig. 2) the blisters reproduce the outline of the dendrite; the central part constitutes a continuous "braid," and the branches consist of point blisters of larger or smaller sizes.

Upon the fusing of a polycrystalline bar (Fig. 3), the network of blisters forms a kind of mosaic of elementary regular meshes corresponding to the individual single-crystal grains in the bar, while the grain boundaries appear sharply in the form of depressions.

When the temperature is raised somewhat, the thickness of the fused layer rises, and overcoming the forces of surface tension, the fused mass runs off, forming a large drop.

This effect may well be of considerable interest in revealing the structural nature of silicon crystals.

CERTAIN CHARACTERISTICS OF THE GROWTH OF SINGLE CRYSTALS OF COMPOUNDS IN THE $A^{III}B^V$ GROUP

M. S. Mirgalovskaya, E. V. Skudnova, and I. A. Strel'nikova

It is well known that the crystallographical conditions of growth exert a large influence on the structure of single-crystal bars of semiconducting materials grown by the Czochralski method.

Up to the present, it is only the growth processes of such elementary semiconductors as Ge and Si that have been subjected to intense study, while semiconducting compounds of the $A^{III}B^V$ group have been considered in less detail. It has been shown that for single crystals of Ge, Si, and compounds of the $A^{III}B^V$ group the most stable direction of growth is the [111]. In Ge and Si, however, the dependence of the crystal growth process on its orientation is associated only with the direction of growth; in compounds of the $A^{III}B^V$ type the growth process is substantially complicated by the effect of the polarity of the [111] directions.

It is known that, in contrast to Ge and Si, which have the diamond lattice, in substances with the ZnS structure any of the crystallographic (111) planes constitutes a combination of two geometric planes, each of which consists of one row of atoms (Fig. 1). The plane formed by B^V atoms is arbitrarily called the B(111) surface, and that formed by A^{III} atoms the A(111).

The atomic model of the A(111) and B(111) surfaces, according to Gatos and Lavine, is shown in Fig. 2.

As first indicated by Gatos et al. [1] and Gatos and Lavine [2] for InSb, the A and B surfaces have substantially different properties; in particular, the chemical activity of the B surface is greater than that of the A surface. Gatos and Lavine also noted that the growth of undoped InSb crystals in the B[111] direction took place more easily than in the A[111] direction.

Hence it was logical to suppose that the characteristics of the (111) planes in the lattices of $A^{III}B^V$-type compounds could lead to different degrees of impurity capture on drawing doped crystals in the B[111] and A[111] directions, and, in particular, to a considerable change in the distribution coefficient of the impurity in the bar.

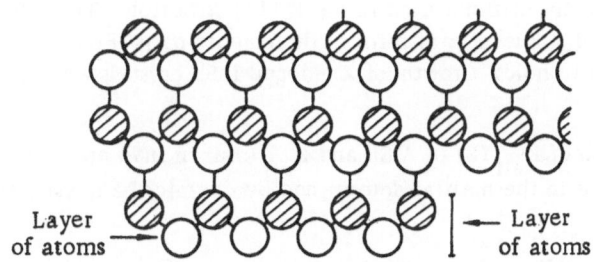

Layer of atoms →   ← Layer of atoms

Fig. 1. Two-dimensional network of (111) planes in the ZnS-type lattice.

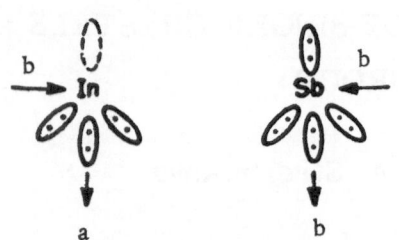

Fig. 2. Atomic models of the A(111) and B(111) surfaces according to Gatos and Lavine.

In view of this, we proceeded to study the effect of the growth polarity on the effective impurity distribution coefficient in $A^{III}B^{V}$, namely, that of S in AlSb and Zn in InSb. First we determined the effect of the polarity of the [111] directions on the growth process of the doped bar.

Experimental Part

The raw materials were mark-B-3 or B-4 Al, mark-In-1000 In, and mark-SU-000 Sb.

The S used for doping contained traces of Mg and As, according to qualitative spectral analysis, and the Zn traces of Al, Cu, Si, and Pb.

The seed crystals were oriented by the Laue method in a camera of the RKSO type in copper radiation to accuracy of ± 2°.

The A(111) and B(111) surfaces were determined by etching. The AlSb samples were etched in H_2O_2: HF: C_2H_5OH = 1:1:8 solution, and the InSb samples in CP-4A (HNO_3: HF: CH_3COOH = 5:3:3) The etchants cited are selective for the A(111) surface, and in 1 to 3 sec, particularly in InSb, etch figures connected with α-60° dislocations appear (Figs. 3 and 4). These etchants do not affect the dislocations on the B surface.

Single crystals of AlSb and InSb were grown by the Czochralski method. The bars were drawn in the B[111] and A[111] directions under optimum conditions, preventing the development of inhomogenities in the crystal. The inhomogeneity of the distribution of S in AlSb was revealed by the infrared microscope, and that of Zn in InSb by electrical measurements. The AlSb bars were drawn under the following conditions: rate of drawing ~0.6 mm/min, seed-rotation rate 23 rpm, crucible-rotation rate ~2 rpm. The InSb bars were grown as follows: rate of drawing ~0.7 mm/min, crystal-rotation rate ~20 rpm, crucible- rotation rate ~13 rpm. The crystallization front, determined by breaking the crystal away from the melt, was slightly concave into the bar.

The impurity concentration in the melt was determined from a sample taken from the melt before drawing the bar, and the impurity concentration in the solid phase corresponded to that of the upper part of the single crystal drawn. The impurity content was determined by chemical analysis: the S in AlSb by the iodometric method, and the Zn in InSb polarographically.

The sulfur and zinc concentrations were determined to accuracies of 0.002 and 0.003% respectively.

It transpired from these experiments that the growth of undoped AlSb single crystals in the A[111] direction was not significantly more difficult than in the B[111] direction. The InSb bars grown in the A[111] direction were, as a rule, only of single-crystal form in their upper parts (Fig. 5). In S-doped AlSb single crystals the same law was found to hold. Growth of Zn-doped InSb crystals was impeded in both directions, namely, A and B[111].

Data on the determination of K_{eff} (S) in AlSb and K_{eff} (Zn) in InSb are presented in Tables 1 and 2, where C_L is the impurity content in the melt as determined by chemical analysis, and C_s is the content in the leading part of the bar.

It follows from Table 1 that K_{eff} (S) in AlSb is less than unity for crystal growth in the A[111] direction, being 0.55 ± 0.063, and for growth in the B[111] direction it is greater than unity, being 1.9 ± 0.18.

It follows from Table 2 that K_{eff} (Zn) in InSb is in all cases greater than unity and equal to 2.1 ± 0.23 for growing single crystals, under the conditions mentioned, in the B[111] direction, and 1.5 ± 0.15 for the A[111] direction.

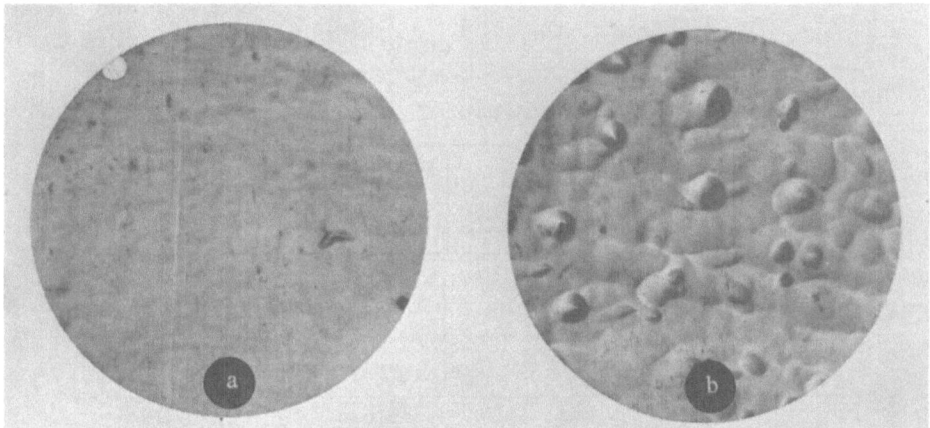

Fig. 3. Microstructure of A(111) and B(111) surfaces of an InSb
bar after etching in CP-4A

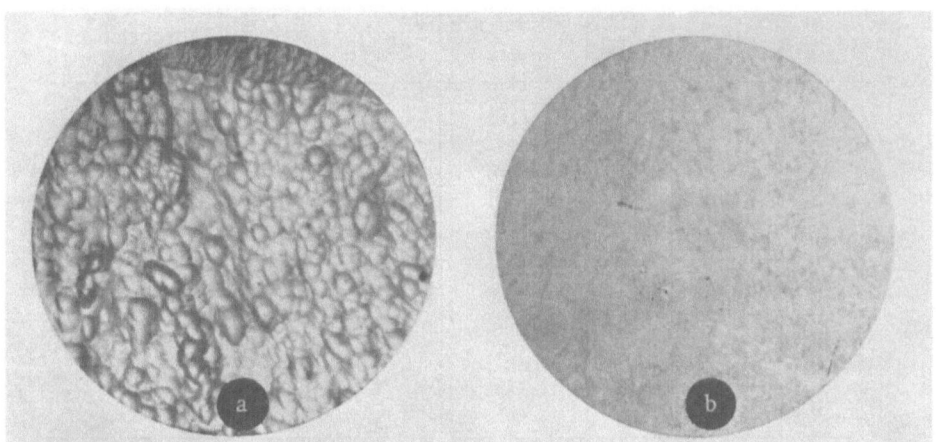

Fig. 4. Microstructure of A(111) and B(111) surfaces of an AlSb
bar after etching.

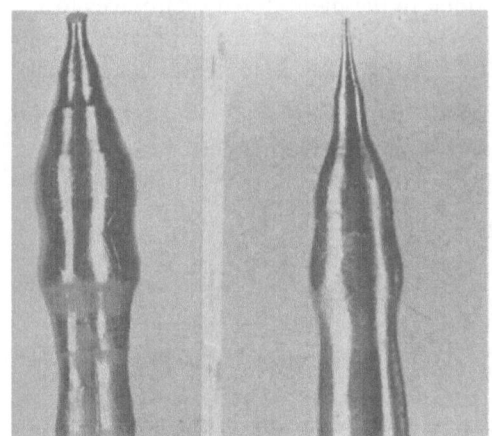

Fig. 5. Undoped InSb single crystals grown in
the A[111] and B[111] directions.

We see from a comparison of the values obtained that the data for K_{eff} differ from one another by amounts exceeding the error in determining either.

Thus it was established that K_{eff} (S) in AlSb is greater in the B[111] direction than in the A[111], for the latter being $K_{eff} < 1$ and for the former $K_{eff} > 1$, whereas K_{eff} (Zn) in InSb is greater for the A[111] direction than for the B[111], exceeding unity in both cases.

Discussion of Results

There is a characteristic kind of growth in crystals of A^{III}B^{V}-type compounds in the [111] direction, both for doped and undoped samples. We have studied the growth characteristics of doped bars by examining single crystals of AlSb and InSb.

TABLE 1. Data on $K_{distrib}$ (S) in AlSb

C_L, wt. %	C_S, wt. %	$K_{eff} = C_S/C_L$	$K_{eff. av}$	Direction of growth
0.050	0.024	0.5		
0.050	0.027	0.5	0.55 ± 0.063	A [111]
0.035	0.022	0.6		
0.060	0.065	(1.1)		
0.030	0.060	2.0		
0.035	0.060	1.7	1.9 ± 0.18	B [111]
0.030	0.060	2.0		

TABLE 2. Data on $K_{distrib}$ (Zn) in InSb

C_L, wt. %	C_S, wt. %	K_{eff}	$K_{eff. av}$	Direction of growth
0.05	0.10	1.6		
0.07	0.14	1.3		
0.015	0.03	1.6		
0.03	0.06	1.5	1.5 ± 0.15	B [111]
0.05	0.11	1.5		
0.02	0.05	1.5		
0.05	0.08	2.0		
0.07	0.09	2.0		
0.04	0.06	2.0		
0.02	0.03	2.0	2.1 ± 0.23	A [111]
0.04	0.06	2.5		
0.04	0.06	2.5		

The results of our investigations show that the $K_{distrib}$ of impurities in the growing crystal depends on the polarity of the [111] direction. One probable explanation of this phenomenon is given below, using the atomic model of Gatos and Lavine [2]. Using this model, we may attempt to explain the growth of the crystals from the melt by supposing that, as the bar grows, groups of atoms formed in the melt attach themselves to the A[111] and B[111] surfaces projecting at the crystallization front. As experiment shows, such groupings arise in the melt as a result of changes in the short-range order structure taking place at temperatures close to the freezing point [3].

It is natural to suppose that the junction takes place in such a way as to preserve the periodicity in the structure of the $A^{III}Sb$ lattice. When the bar is growing in the A[111] direction, it is clear that bonds of $A^{III}_{sol} - (Sb-A^{III} \ldots)_{melt}$ are mainly formed, while for the B[111] direction we have mainly $Sb_{sol} - (A^{III} - Sb)_{melt}$ bonds.

Since the growth of undoped bars in the A[111] direction is more difficult than in the B[111] direction, clearly the formation of $A^{III}_{sol} - (Sb-A^{III})_{melt}$ bonds is more difficult than that of $Sb_{sol} - (A^{III} - Sb)_{melt}$ bonds [2].

Doping the AlSb melt with S does not introduce any special changes into the growth of the bar in the B[111] direction, since the same bond $Sb_{sol} - [Al-Sb (S)]_{melt}$ is realized. In this case the amount of sulfur taken up by the growing crystal is determined by the concentration of $(Al-S \ldots)_{melt}$ groupings at the crystallization front. When the single crystals are grown from a sulfur-doped melt in the A[111] direction, the bonds formed are mainly $Al_{sol}-[S-Al...]_{melt}$. The attachment of $[S-Al...]_{melt}$ groups must lead to an impediment in the growth process, since the organization of $Al_{sol}-[S-Al...]_{melt}$ bonds is more difficult than that of $Al_{sol}-[Sb-Al...]_{melt}$ bonds, since sulfur is an element with a larger valence and a smaller tetrahederal radius than Sb (1.04 and 1.36 A, respectively). In this case the growth of the bar evidently involves the preferential attachment of the $[Sb-Al...]_{melt}$ groupings. Clearly this also explains the fact that the $K_{eff}(S)$ in AlSb for growth in the A[111] direction is considerably smaller than for growth in the B[111] direction. It is probably for this reason also that the growth of AlSb crystals doped with sulfur takes place more easily in the B[111] direction than the A[111]. An analogous explanation may be given for the process of doping InSb with zinc.

In growth of the bar in the A[111] direction doping with zinc introduces no changes, since the same bond, $In_{sol}-[Sb-In(Zn)...]_{melt}$, is realized. With growth of crystals from a zinc-doped melt in the B[111] direction, however, the process involves the formation of $Sb_{sol}-[In(Zn)-Sb...]_{melt}$ bonds. The attachment of $(Zn-Sb...)_{melt}$ groups must lead to an impediment in the growth of the bar, since Zn is an element with lower valence and a smaller covalent radius than indium ($R_{Zn} = 1.31$ A, $R_{In} = 1.44$ A).

In this case the growth of the bar clearly involves preferential attachment of $[In-Sb...]_{melt}$ groups.

Clearly this also is responsible for the fact that $K_{eff}(Zn)$ in InSb is somewhat smaller for growth in the B[111] direction than for the A[111]. This probably also explains the slight difference in the growth of zinc-doped crystals in the A[111] and B[111] directions.

It must be noted that the explanation given above is only approximate. Nevertheless, it has been shown experimentally that, for $A^{III}B^V$ compounds, the K_{eff} of impurities depends on the polarity of the [111] directions of growth. However, its value in every specific case is obviously governed by the nature and quantity of the doping component.

It has thus been established that the character of the growth of doped single crystals of the $A^{III}B^V$ type depends on the polarity of the [111] directions, which affects both the growth process of doped material and the value of $K_{distrib}$. From this follows the important practical conclusion that, by using the effect of the polarity of the growth direction we may vary the value of $K_{distrib}$ as desired (in specific cases from values > 1 to values < 1).

Conclusions

1. The effect of the polarity of the A[111] and B[111] directions on the growth process of InSb and AlSb single crystals has been determined.

2. The effect of the polarity of the A[111] and B[111] growth directions on the value of K_{eff} for certain impurities in AlSb and InSb has been established.

3. For crystal growth in the A[111] and B[111] directions, K_{eff} takes the following values: for S in AlSb, 0.55 ± 0.068 and 1.9 ± 0.18, and for Zn in InSb 2.1 ± 0.13 and 1.5 ± 0.15, respectively.

Literature Cited

1. H. C. Gatos, P. L. Moody, and M. C. Lavine, J. Appl. Phys. 31(1):212 (1960).
2. H. C. Gatos and M. C. Lavine, J. Phys. Chem. Solids, (14):169 (1960).
3. V. M. Glazov, Izv. Akad. Nauk SSSR, Otd. Tekhn. Nauk, Metal. i Toplivo, No. 5: 190 (1960).

LAWS OF POLYMORPHIC TRANSFORMATIONS OF ELEMENTS IN CONNECTION WITH THEIR ELECTRONIC STRUCTURES

V. K. Grigorovich

Polymorphic transformations of elements which take place with variations in temperature, as is the case with phase transformations from one state of aggregation to another, are distinguished by the temperature-dependence of the thermodynamic potentials of the phases (Fig. 1). The phase stable at a given temperature is that with the lower value of potential z, and the equilibrium temperature of the phase transformation corresponds to equal potentials of both phases, $Z_\alpha = Z_\beta$.

The form of the thermodynamic potential curves is determined by the variation in the specific heat of the substance with increasing temperature resulting from the change in the amplitude and frequency of atomic vibrations (atomic specific heat) and the increase in the energy of the electrons with increasing temperature (electronic specific heat). With increasing temperature, the second component rises and may prove substantial. A considerable change in specific heat takes place on spin coupling of the electrons without change of crystal structure (magnetic transformation of iron, cobalt, and other ferromagnetics). Transformations involving the state of aggregation (melting or evaporation), which take place as a result of the local rupture of interatomic bonds, or their complete disruption on atomization, have a very different nature from polymorphic transformations, in which the mutual bond between atoms is partly broken just at the moment of transformation and then re-established, ultimately leading to a change in their mutual dispositions, as characterized by a change of coordination number and interatomic distances. We may well imagine that in the polymorphism of elements and compounds expressed in geometrical changes of crystal structure, an essential part must be played not only by the energetic characteristics of the lattice but also by the structure of the outer electron shells of the atoms and ions, namely, the electronic configuration, dimensions, shape, and type of interaction of the outer electron shells determining the form of the atoms and ions in the lattice. Thus consideration of the electron density in the structures of a number of elements and compounds indicates that spherical symmetry is absent from the component atoms and ions, while in many cases there are directional bonds [1].

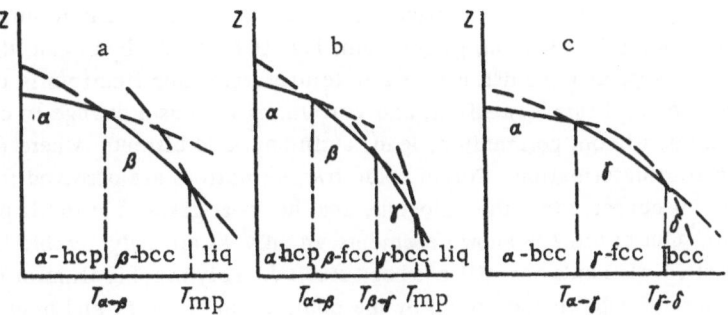

Fig. 1. Variation of the Gibbs free energies of the crystalline forms of a) titanium, b) lanthanum, and c) iron with increasing temperature.

The crystal structure of many metals becomes thermodynamically unstable as temperature rises, and is replaced by stabler modifications. These reversible allotropic transformations have recently been discovered in many metals earlier considered monomorphic, for example, lithium, sodium, beryllium, scandium, yttrium, most of the lanthanides, actinium, thorium, and other actinides. It is not impossible for polymorphic transformations at high temperatures as well as those even below 0°C to be found in other metals at present considered monomorphic.

As a result of intensive experimental work, the polymorphism of metals is now to a great extent understood. Polymorphic transformations, such as the $\alpha \to \gamma$ -transformation of iron, discovered by Chernov, from the basis of the entire theory and practice of the heat treatment of steel and the theory of iron-base alloys. Also of collosal practical importance are the polymorphic transformations of titanium, zirconium, uranium, and certain other metals. However, the laws of polymorphic transformations, in the sense of the sequence of the changing forms as temperature rises, have not been strictly established, and the cause of polymorphism in metals remains unexplained.

Savitskii has made a detailed study of metals from the point of view of polymorphic transformations with changes in temperature; he has studied their mechanical properties, noted a number of general laws of polymorphism and the conditions for changes in mechanical properties, and has come to the conclusion that the most high-temperature forms of polymorphic metals "have cubic and in the overwhelming majority of cases face-centered structures" [2].

Many attempts have been made to correlate the structure of metals with the electronic structure of their atoms on the basis of resonance theory [3—5, etc.], but these have not been entirely successful, if only because the explanation of the body-centered cubic structure of the transitional metals (just as cubic hexagonal close-packing) based on the assumption that the s-, p-, and d-electrons hybridize or that directional bonds are formed, as proposed for transitional metals with d-electrons, is invalid for alkali and alkaline earth metals, which have no p- and d-valence electrons at all, but nevertheless have body-centered cubic structures. The existence of body-centered cubic structures in different metals must clearly result from a common cause, namely, the similar structures of the outer shells in their ions.

Let us consider the crystal structures and polymorphic transformations of the elements in connection with their position in the periodic system in order to establish and classify general laws of transformation.

At least half of the elements possess the property of polymorphism (Table 1). Temperature polymorphism has not yet been observed in the following elements: potassium and rubidium (Group Ia); magnesium (Group IIa); europium, erbium, thulium, (lanthanides), protactinium, vanadium, niobium, tantalum (Group V); chromium, molybdenum, tungsten (Group VI); technecium, rhenium (Group VII); ruthenium, osmium, iridium, nickel, palladium, platinum (Group VIII); copper, silver, gold (Group Ib); zinc, cadmium, mercury (Group IIb); aluminum, gallium, indium (Group III); silicon, germanium, lead (Group IV); hydrogen, the halogens, and the inert gases. There are data to support the existence of low-temperature modifications in cesium and vanadium, and high-temperature forms in beryllium, scandium, and yttrium, as well as a change in coordination number on melting in gallium, indium, silicon, germanium, lead, antimony, and bismuth, where melting apparently coincides with a polymorphic transformation. Polymorphic transformations are observed in elements of all groups except the subgroups of copper, zinc, the halogens, and the inert gases, and in all periods except the first (hydrogen and helium). Out of the 103 known elements, in only 47 are polymorphic transformations not found, while the structure of 11 elements has still not been studied. Polymorphic transformations also take place in metals of the main subgroups characterized by the completion of the s- and p-shells, for example, in lithium, sodium, calcium, strontium, barium, thallium, and tin (Table 2). They are inherent in many non-metals with p-shells which are filling, such as boron, carbon, nitrogen, phosphorus, arsenic, antimony, sulfur, selenium, tellurium, and polonium, and their transformations characterize metals with nearly-finished d-shells, namely, metals of the scandium and titanium subgroups, as well as manganese, iron, and cobalt. Finally, almost all the elements with filling 4f- and 5f-shells (lanthanides and actinides) are polymorphic metals. Among these, modifications have not yet been found only in europium, erbium, thulium, and protactinium.

TABLE 1. Polymorphism of the Elements in the Periodic System

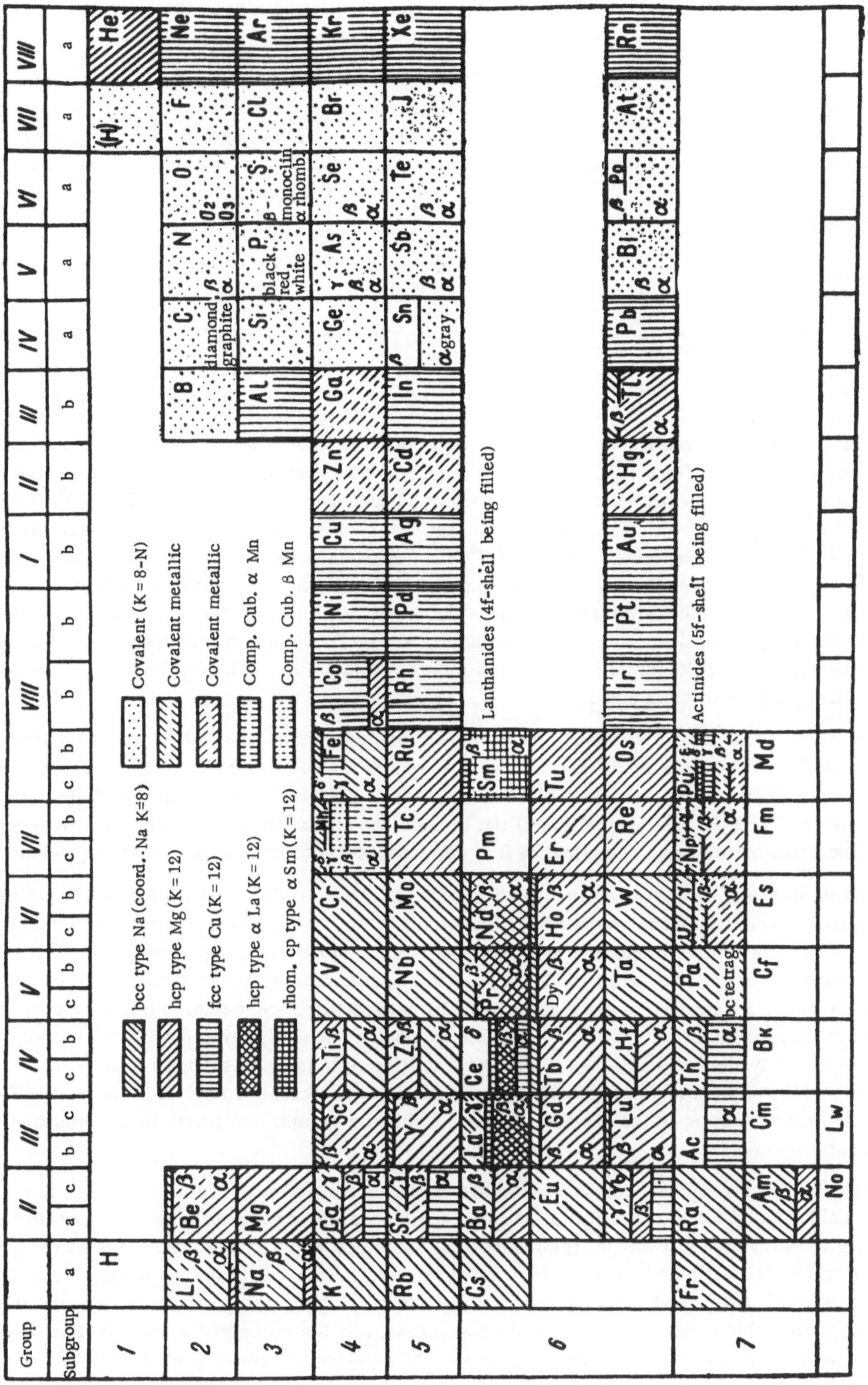

Legend:

- Covalent (K = 8-N)
- Covalent metallic
- Covalent metallic
- Comp. Cub. α Mn
- Comp. Cub. β Mn

- bcc type Na (coord. Na K=8)
- hcp type Mg (K=12)
- fcc type Cu (K=12)
- hcp type α La (K=12)
- rhom. cp type α Sm (K=12)

Group	I	II	III		IV	V	VI	VII	VIII			
Subgroup	a	a	b	c	c	c	c	c	c	b	b	a

Row 1: H; He

Row 2: Li β/α; Be β/α; B; C (diamond β, graphite α); N; O (O₂, O₃); F; Ne

Row 3: Na β; Mg; Al; Si; P (black P, red, white); S (β monoclin, α rhomb); Cl; Ar

Row 4: K; Ca γ/β/α; Sc β/α; Ti β/α; V; Cr β/α; Mn (β, α); Fe β/α; Co β/α; Ni; Cu; Zn; Ga; Ge; As γ/β/α; Se β/α; Br; Kr

Row 5: Rb; Sr β/α; Y β/α; Zr β/α; Nb; Mo; Tc; Ru; Rh; Pd; Ag; Cd; In; Sn (α gray, β); Sb β/α; Te β/α; J; Xe

Row 6: Cs; Ba β/α; La β/α; Ce δ/γ/β/α; Pr β/α; Nd β/α; Pm; Sm β/α; Eu; Gd β/α; Tb β/α; Dy β/α; Ho β/α; Er; Tu; Lu β/α; Hf β/α; Ta; W; Re; Os; Ir; Pt; Au; Hg; Tl β/α; Pb; Bi β/α; Po β/α; At; Rn

Lanthanides (4f-shell being filled)

Row 7: Fr; Ra; Ac; Th β/α; Pa (bc tetrag); U β/α; Np; Pu ...; Am β/α; Cm; Bk; Cf; Es; Fm; Md; No; Lw

Actinides (5f-shell being filled)

TABLE 2. Prevalence of Polymorphic Transformations in the Elements

Elements	Completion of electron shell	Polymorphism observed, elements %	Polymorphism not observed elements %	Structure not studied, elements %	Total numbers of elements, elements %
Main subgroups (a)	s^2p^6	$\dfrac{19}{42}$	$\dfrac{23}{51}$	$\dfrac{2}{7}$	$\dfrac{44}{100}$
d-transitional metals (b)	d^{10}	$\dfrac{11}{35.5}$	$\dfrac{20}{64.5}$	0	$\dfrac{31}{100}$
Lanthanides and actinides (c)	f^{14}	$\dfrac{18}{64.4}$	$\dfrac{1}{3.6}$	$\dfrac{9}{32}$	$\dfrac{28}{100}$
Total	—	$\dfrac{48}{46.5}$	$\dfrac{44}{42.6}$	$\dfrac{11}{10.8}$	$\dfrac{103}{100}$

Thus polymorphic transformations are inherent both metals and nonmetals, as well as in elements with filling s- and p-, d-, and f-shells (Table 2). We may just note that many elements with completely filled outer shells do not tend toward polymorphism, namely, inert gases (filled shells $1s^2$ and s^2p^6), and elements with filled outer d^{10}-shells over which lie one or two electrons, such as zinc, cadmium, and mercury (electron configuration $d^{10}s^2$), and copper, silver, and gold ($d^{10}s^1$). Remaining without modification are those metals with almost-filled d-shells — nickel, palladium, platinum, iridium, ruthenium, osmium, and radium (except iron and cobalt) — as well as the transitional metals of Groups V, VI, and VII (except manganese). Polymorphic modifications are not found in lanthanides with configuration f^6 (europium) and with an almost-filled f-shell (erbium and thulium). Polymorphic transformations are not found in the halogens, the structures of which comprise strong diatomic molecules.

The crystal structures of various forms of metals and nonmetals are shown conventionally in Table 1; the areas occupied by the various modifications indicate approximately the temperature range in which they exist in fractions of the absolute temperature of the corresponding melting points. The transformation temperatures are indicated in Table 3. The disposition of the lanthanides and actinides in groups in periods 6 and 7 between the alkaline earth metals and the analogs of iron corresponds to the similarity of their crystal structures.

Moving from the nonmetals to the alkali metals, i.e., from right to left in the periods of Mendeleev's table, we find a transition from the molecular structures of the inert gases and halogens to the covalent structures of nonmetallic elements of the oxygen, nitrogen, and carbon groups, and then to the covalent— metallic structures of gallium, white tin, zinc, cadmium, and mercury with indications of directional bonds. We then encounter a transition to the typical metallic lattice, the face-centered cubic (copper, nickel, and cobalt subgroups), followed by the hexagonal close-packed, typical of many transitional metals and lanthanides, and finally the body-centered structure typical of transitional metals of Groups III through VI, alkaline earths, and alkali metals. An analogous tendency, associated with the increase in metallic properties, is also noted on moving from top to bottom in the nitrogen, carbon, and boron groups, and partly in the groups of alkali and alkaline earth metals.

Let us consider a typical sequence of change in the crystal forms of elements at high temperatures. We see from Tables 1 and 3 that, as temperature rises, some covalent crystals pass into the metallic state; for example, gray tin, which is a brittle, semiconducting substance with the diamond structure (K = 4), transforms into white tin, which has a body-centered tetragonal structure with a cordination number approximately equal to 6, high ductility, and metallic conductivity. Increasing the temperature leads to the transformation of the composite cubic structures of α- and β- manganese, and the composite structures of α- and β- uranium and neptunium and α-, β-, and γ-plutonium, which are characterized by reduced ductility and the presence of directional bonds, into a centered cubic structure typical for metals; in manganese and plutonium this transformation is preceded by a transformation to a cubic close-packed form. For the majority of polymorphic

TABLE 3. Crystal Modifications and Transformations of Metals [6—8]

Metal	α	Transformation temperature, °C	β	Transformation temperature, °C	τ	Transformation temperature, °C	δ	Transformation temperature, °C	δ'
Lithium	hcp	−195	bcc	180.5	Liquid				
Sodium	hcp	−268	bcc	97.8	Liquid				
Beryllium	hcp	1250	bcc	1282	Liquid				
Calcium	fcc	250	fcc	464	bcc	850	Liquid		
Strontium	fcc	248	fcc	589	bcc	770	Liquid		
Scandium	hcp	1335	bcc	1539	Liquid				
Yttrium	hcp	1495	bcc	1509	Liquid				
Lanthanum	hcp	310	fcc	868	bcc	920	Liquid		
Cerium	fcc	−150	hcp	− 10	fcc	725	bcc		
Praseo-dymium	hpc	788	bcc	335	Liquid				
Neodymium	hpc	868	bcc	1024	Liquid				
Samarium	rhombohedral	917	bcc	1072	Liquid				
Gadolinium	hcp	1262	bcc	1312	Liquid				
Terbium	hpc	1316	bcc	1356	Liquid				
Dysprosium	hpc	1392	bcc	1407	Liquid				
Holmium	hpc	1442	bcc	1461	Liquid				
Ytterbium	fcc	798	bcc	824	Liquid				
Lutecium	hpc	1407	bcc	1652	Liquid				
Thorium	fcc	1400	bcc	1827	Liquid				
Uranium	orthorhombic	660	tetrag.	770	bcc	1133	Liquid		
Neptunium	rhombohedral	278	tetrag.	570	bcc	637	Liquid		
Plutonium	monoclinic	119	bc monoclin.	218	comp.	310	fcc	450	fc tetrag. (472) bcc (640) Liquid
Americium	hexag.	—	bcc	1200	Liquid.				
Titanium	hpc	882	bcc	1700	Liquid				
Zirconium	hpc	867	bcc	1860	Liquid				
Hafnium	hpc	1975	bcc	2230	Liquid				
Manganese	comp. cub.	727.2	comp. cub.	1094	fcc	1134	bcc	1244	Liquid
Iron	bcc	768	bcc	911	fcc	1390	bcc	1540	Liquid
Cobalt	hpc	470	fcc	1495	Liquid				
Thallium	hpc	234	bcc	302.5	Liquid				
Tin	diamond	13.2	tetrag.	232	Liquid				
Polonium	orthorhombic	18—54	rhombohedral	254					

Notes: hcp = hexagonal close-packed structure (K = 6 + 6); bcc = body-centered cubic structure (K = 8); fcc = face-centered cubic structure (K = 12); diamond = structure of the diamond type (K = 4).

TABLE 4. Ratio c/a for Metallic Structures at 20°C [6—8]

Metal	Type of structure	c/a	Deviation of the form of the ion from spherical along c axis
Beryllium	Hexagonal close-packed (hpc) compressed	1.5677	0.960
Holmium		1.571	0.961
Erbium		1.571	0.961
Thulium		1.572	0.962
Yttrium		1.5722—1.5712	0.962
Dysprosium		1.574	0.964
Osmium		1.5790—1.584	0.9576
Terbium		1.583	0.967
Ruthenium		1.5824	0.968
α-Titanium		1.5873	0.971
α-Hafnium		1.5881—1.5811	0.972
Gadolinium		1.591	0.972
Scandium		1.5917	0.974
α-Zirconium		1.5992—1.5931	0.978
α-Thallium		1.5984	0.979
Technetium		1.604	1.982
Lanthanum		1.607	0.984
Praseodymium		1.610	0.986
Samarium		1.610	0.986
Rhenium		1.6148	0.988
Cerium		1.618	0.990
Neodymium		1.620	0.991
α-Cobalt*		1.6232—1.6322	0.993
Magnesium		1.6235	0.993
Hydrogen		1.63	0.998
Metals isomorphic with copper	Cubic close-packed (fcc) along c axis	1.63333...	1.0000
Metals isomorphic with α-Fe and W	Body-centered cubic	1.63333...	1.0000
Protactinium	Body-centered tetragonal	0.825†	0.825
δ'-Plutonium	Face-centered tetragonal	0.954†	0.954
Indium		1.07586†	1.07586
Sodium (at 5°K)	Hexagonal close-packed (hpc) drawn out along c axis	1.634	1.001
Helium		1.635	1.001
β-Strontium		1.636	1.001
Ytterbium		1.637	1.002
β-Calcium		1.638	1.002
Lithium (at 78°K)		1.637	1.002
Zinc		1.8563	1.136
Cadmium		1.8856	1.156
Mercury (at -46°K)	Rhombohedral (distorted hexagonal)	1.937	1.185

*Close-packed hexagonal and cubic forms coexist at low temperatures.
†Ratio of height to side of base (b/a).

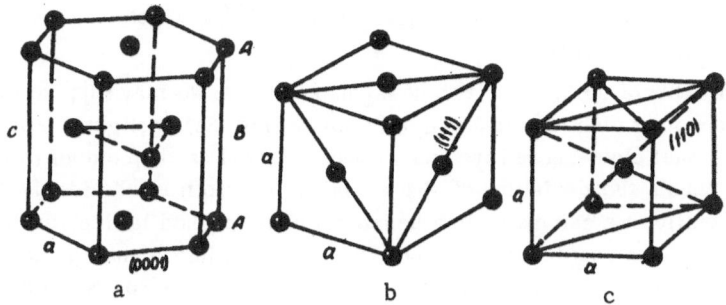

Fig. 2. Scheme for the calculation of the ratio of twice the dis--
tance between the most closely packed planes (c) to the shortest
distance between atoms in the plane (a) in typical metal structures;
a) hexagonal close-packed ideal c/a = $\sqrt{8/3}$, actual c/a ≠ $\sqrt{8/3}$);
b) face-centered cubic c/a = $4a/(\sqrt{3}\cdot a\sqrt{2}) = \sqrt{8/3}$; c) body-centered
cubic c/a = $2a\sqrt{2}/a\sqrt{3} = \sqrt{8/3}$.

metals, the low-temperature α-form is the cubic or hexagonal close-packed one, changing to body-centered
cubic as the temperature rises (lithium, sodium, beryllium, calcium, strontium, barium, scandium, yttrium,
lanthanum, actinium, and the majority of the lanthanides, as well as titanium, zirconium, hafnium, thallium,
etc. —29 metals in all). In calcium, strontium, and apparently ytterbium, as temperature rises there is a
transformation from the cubic close-packed α-modification to the hexagonal close-packed β-modification,
and then on to the body-centered form. Sometimes, between the hexagonal close-packed and body-centered
cubic structures, there exists a region in which face-centered cubic packing is stable (lanthanum, cerium). In
a number of cases, as temperature rises, there is a transformation of the face-centered cubic structure to body-
centered (thorium, $\gamma \rightarrow \delta$ transformation of manganese, $\gamma \rightarrow \delta$ of iron) or of hexagonal into face-centered
cubic (cobalt). The most typical sequence of changes in structure as temperature rises is a transition from covalent
and covalent—metallic structures with directional bonds and low coordination numbers to close-packed hexagonal
(K = 6 + 6) or cubic (K = 12), and finally to body-centered cubic (K = 8), which as a rule, is the highest-temperature
form.

In many elements, individual members of this sequence do not exist, but only in iron as temperature
rises is there a transformation from the body-centered form to the close-packed cubic γ-form, although the
$\gamma \rightarrow \delta$ transformation of iron corresponds to the typical sequence discussed above. From the thermodynamic
point of view, the transformation from one modification to another as temperature rises is caused by the intersection of the Gibbs free energy curves (z) for the respective phases, the stability of dense packing (K = 12) at
low temperatures and the transition to the less dense (K = 8) body-centered cubic at high temperatures, corresponding to the increased entropy of the latter (see Fig. 1a, b). The Gibbs-function curves shown in Fig. 1c
show that the anomalous behavior of iron may be explained by the fact that the curve for γ-Fe has a smaller
curvature than that for α-Fe and intersects the latter at two points.

Let us clarify the possible causes of polymorphic transformations. First of all, we note that all metals
with hexagonal close-packed structure have the axial ratio c/a either smaller or larger than the ideal value
corresponding to the dense packing of spheres, the discrepancy occurring in the second place of decimals
(Table 4), whereas the accuracy of the determining parameters a and c by X-ray structural analysis extends to
the fourth and fifth places. The most "ideal" lattice of magnesium is characterized by the ratio c/a = 1.6235.
For the majority of transitional metals and lanthanides with outer d- and f-electrons and hexagonal structures,
c/a < 1.6333..., but for many metals of the main subgroups with outer s-electrons c/a > 1.6333... .

In a hexagonal close-packed lattice of the Mg type with layer alternation ABAB, c represents twice the
distance between neighboring close-packed basal layers A and B, and the shortest distance between atoms in

such a layer equals a. The diameter of the atom equals $\sqrt{a^2/3 + c^2/4}$, which for close-packed spheres leads to the ratio $c/a = \sqrt{8/3} = 1.6333...$ (Fig. 2a).

From the nonideal value of the ratio c/a in hexagonal lattices, we may suppose that these are made up of atoms or ions with their outer electron clouds slightly oblate (Fig. 3a) or slightly prolate (Fig. 3b). In other words, the atoms in hexagonal close-packed systems may be pictured as ellipsoids of rotation in which the ratio of the semiaxes a and b equals the deviation of the actual c/a ratio from the ideal ($\sqrt{8/3}$) in sign magnitude. From such ellipsoids of rotation we can make up a hexatonal close-packed lattice with coordination number 6, but not cubic close-packing with coordination number 12. Actually, just as there is not a single hexagonal close-packed lattice with a ratio c/a exactly equal to 1.6333... (Table 4), so also in metals, as a rule we find no tetragonally distorted cubic packings. These appear only as exceptions in indium and δ'-polutonium.

Let us find the c/a ratio for close face-centered cubic packing. Here c must be taken as twice the distance between neighboring close-packed octahedral (111) planes, and a as the shortest distance between the atoms in these planes, i.e., the diameter of an atom.[*] From Fig. 2b we have $d = a\sqrt{2}/2$, where a is the lattice parameter, and $c = 2a/\sqrt{3}$, whence $c/a = 2a \cdot 2/\sqrt{3} \cdot a\sqrt{2} = \sqrt{8/3} = 1.6333...$. Thus the c/a ratio for the face-centered cubic lattice strictly equals the ideal ($\sqrt{8}/3$). Hence we may assume that the cubic close-packed structure is composed of atoms with spherical outer electron shells.

In metals with hexagonal close-packed structure there is a change in c/a with rising temperature; this may be because the atoms approach spherical form (Fig. 4, Table 4). In magnesium the ratio c/a changes very little with rising temperature, while in beryllium and thallium it falls (Fig. 4). Since at low temperatures the c/a ratio of thallium approaches 1.6333..., it is not possible that thallium may have a cubic close-packed modification at high pressures and at temperatures below -250°. In calcium, strontium, and ytterbium, the low-temperature α-form has cubic close-packing, and at higher temperatures there appears a hexagonal close-packed form, which in calcium and ytterbium is apparently due to impurities [6]. The fact that the c/a ratio in the hexagonal forms of these three metals is close to the ideal value results in their being metastable.

The hexagonal forms of all the transitional metals and lanthanides, except zinc and cadmium, which have cells extended anomalously along the c axis, are characterized by $c/a < 1.6333...$, the ratio rising with increasing temperature (Fig. 4). As the c/a ratio approaches the ideal value (350 to 400° in cobalt, 310° in lanthanum, and -10° in cerium), the hexagonal close-packed structure of the Mg or α-La type transforms into cubic close-packed. In the majority of cases, before the c/a ratio becomes close enough to 1.6333... , the hexagonal close-packed structure transforms into body-centered cubic. This sort of polymorphism, most typical for metals takes place in the $\alpha \rightarrow \beta$ transformations of scandium, yttrium, praseodymium, neodymium, gadolinium, terbium, dysporsium, holmium, lutecium, titanium, zirconium, hafnium, and thallium. A transformation of face-centered into body-centered cubic structure characterizes the αTh $\rightarrow \beta$ Th, β La $\rightarrow \gamma$ La, γ Ce $\rightarrow \delta$Ce, γ Mn $\rightarrow \delta$ Mn, γ Fe $\rightarrow \delta$Fe, and δ'Pu $\rightarrow \varepsilon$ Pu.

Thus the transformation of hexagonal close-packing into cubic close-packing may arise when the ellipsodial form of the outer electron cloud belonging to metal lattice ions approaches spherical as the temperature changes; it is apparently not connected with any radical change in the configuration of the outer electron cloud as a result of a change in the degree of ionization.

For the transformation of (hexagonal or cubic) close-packing into body-centered cubic, the rise in temperature is accompanied not only by an increase in the energy of thermal vibrations of the atoms, but also by an increase in the energy of the outer electrons. In a metal lattice, where the outer electrons of the positive ions are already excited, owing to the perturbing action of neighboring atoms, a comparatively small rise in temperature may be sufficient for the removal of some outer electrons, in particular, the second s-electrons of alkaline earth metals. Then the orthogonal p^6-shell is uncovered and the centered cubic γ-modification of calcium, strontium, and barium appears. The removal of the fourth valence d-electron in titanium, zirconium, and hafnium as the temperature rises also leads to the formation of ions with an outer p^6 configuration and hence to

---

[*]$h/a = 1$ for fcc.

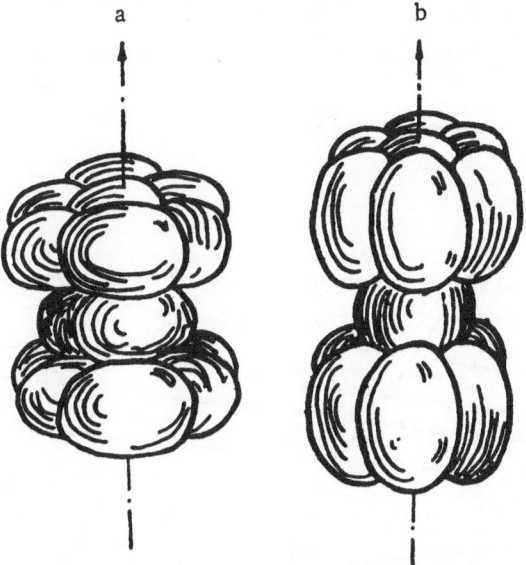

Fig. 3. Hexagonal close-packing of ellipsoidal ions: a) c/a < 1.6333...(all the transitional metals and lanthanides with hcp structure, as well as beryllium, magnesium, and thallium) ; b) c/a > 1.6333... (elements of the main groups: α-Li, α-Na,β-Ca, β-Sr, He, Zn, Cd).

TABLE 5. Variation of the c/a Ratio with Increasing Temperature and Expansion Coefficients of the Transitional Metals with Hexagonal Close-Packed Structure [6, 7]

Metal	Value of c/a		t, °C	Transformation temperature, °C	Type of transformation	Expansion coefficient $\alpha \cdot 10^6$ along axes	
	at 20°C	at t, °C				⊥c	‖c
Titanium	1.5874	1.591	882	882.5	hcp→bcc	11.03	13.37
Zirconium	1.5922	1.59985	811	867	hcp→bcc	5.64	6.39
Hafnium	1.581	—	—	1975	hcp→bcc	—	—
Ruthenium	1.5835	1.5868	600	—	—	7.2	11.0
Osmium	1.5790	1.5810	600	—	—	4.9	7.1
Rhenium	1.615	—	—	—	—	4.67	12.45
Zinc	1.852	1.883	410	—	—	—	—
Cadmium	1.888	1.904	280	—	—	2.21	4.83
Lanthanum	1.607	1.619	293	310	hcp→fcc	—	—
Cerium	1.618	1.618	−10	−10	hcp→fcc	—	—
Praseodymium	1.61	1.619	449	798	hcp→bcc	4.5	11.3
Neodymium	1.61	1.621	696	868	hcp→bcc	6.5	11.7
Samarium	1.61	—	—	917	rhomb.→bcc	—	—
Gadolinium	1.591	1.595	644	1263	hcp→bcc	6.3	13.0
Terbium	1.583	1.595	852	1316	hcp→bcc	9.1	17.9
Dysprosium	1.574	1.590	685	1392	hcp→bcc	4.7	20.3
Holmium	1.571	1.586	708	1442	hcp→bcc	4.9	18.4
Erbium	1.571	1.587	917	—	—	7.5	19.5
Thulium	1.572	1.589	853	—	—	7.5	19.7
Ytterbium	—	1.637	268	—	hcp→bcc	—	—
Lutecium	1.583	1.596	956	1407	hcp→fcc	10.2	16.9
Scandium	1.5917	1.602	1009	1335	hcp→bcc	9.3	15.2
Yttrium	1.5722	1.591	897	1495	hcp→bcc	6.2	19.7

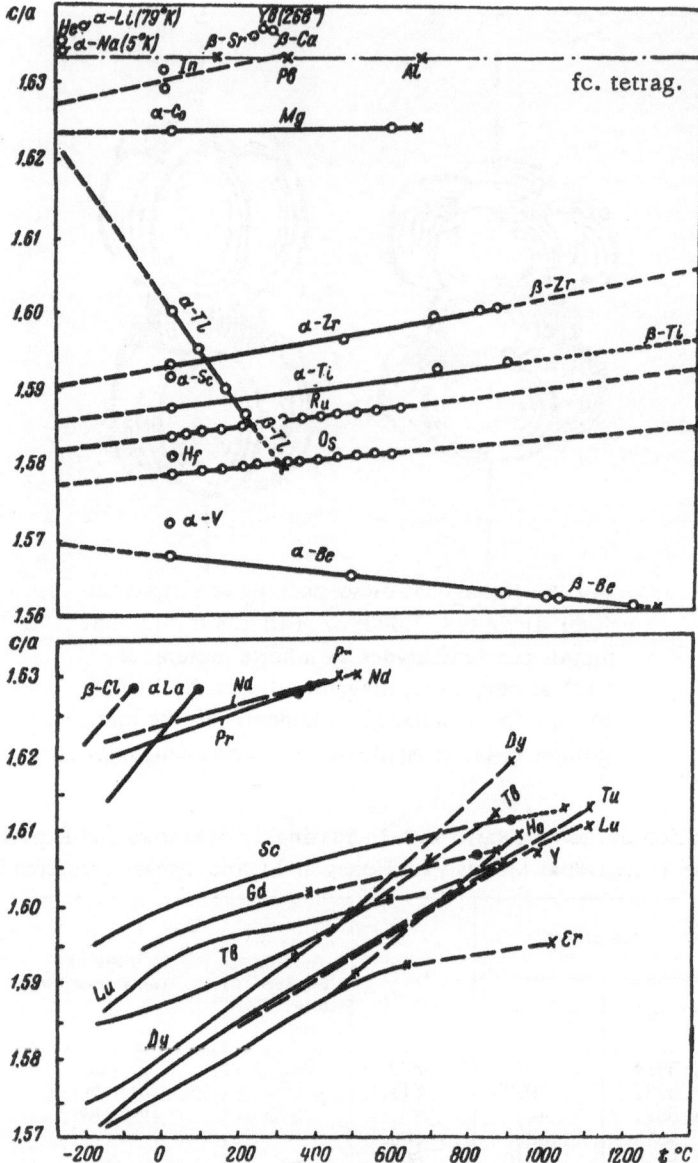

Fig. 4. Variation in ratio c/a with increasing temperature for metals with hexagonal close-packed structure: Metals of the main groups [1⁻3]; transitional metals [4⁻7]; lanthanides [6, 7]: — ○— hpc . . . . . . bcc; — ×— melting point.

the appearance of the body-centered high-temperature β-form. The removal of three valence electrons from thallium leads to the transformation of its hexagonal close-packed α-form into the body-centered cubic β-form. As a result of thermal ionization of the third valence d-electron, ions with the outer p^6 configuration are formed, and the high-temperature body-centered modification appears in scandium, yttrium, lanthanum, and (apparently) actinium. The uncovering of the p^6 configuration when the two s-electrons and one of the two f-electrons pass into the electron gas leads to the appearance of the body-centered cubic forms immediately before melting in cerium, praseodymium, neodymium,samarium, gadolinium, terbium, dysprosium, holmium, ytterbium, and lutecium. For the same reasons, we may expect body-centered forms to develop near the melting point in promethium, erbium, and thulium, and also low-temperature close-packed forms in europium (at high pressures below 0°C) [9].

The formation of body-centered cubic forms at temperatures close to the melting point is also characteristic of all the actinides. The removal of all the $6d^2 7s^2$ electrons leads to the uncovering of a $6p^6$ shell in thorium and the transformation of its face-centered cubic α-form into the body-centered cubic β-structure. As a result of the removal of the outer $6d^1 7s^2$ electrons from uranium and neptunium on increasing the temperature, their covalent—metallic α- and β-forms transform into the body-centered cubic γ-structure, which is also caused by the exposure of the $6p^6$ shell. Finally, as the temperature rises, there is a transition from the composite covalent—metallic α-, β-, and γ-structures of plutonium to the face-centered cubic and tetragonal δ and δ'-modifications, and then, owing to the exposure of the $6p^6$ shell, the highest-temperature body-centered ε-form appears. It is interesting that the high-temperature polymorphism of plutonium with electron configuration $5f^6 6p^6 7s^2$ has features in common with the $\gamma \rightarrow \delta$ transformation of iron (electron configuration $3p^6 3d^6 4s^2$). Body-centered cubic structures apply to all the transitional metals of the fifth and sixth groups, in which the outer p^6 shells are uncovered upon removal of all the valence electrons. The idea that, in Group I to VI metals of the large periods, the number of bond electrons is equal to the higher value of the valence or the Group number was introduced by Pauling [3] and Hume—Rothery et al. [4].

Conclusions

1. As temperature rises, polymorphic metals show the following sequence of changes in form: composite covalent—metallic structures with indications of directional bonds, hexagonal close-packed or cubic close-packed (face-centered cubic), and body-centered cubic. In particular metals, however, some of the forms in this sequence may not exist. The only exception to this rule is the $\alpha \rightarrow \gamma$ transformation of iron, whereas the $\gamma \rightarrow \delta$ transformation of iron obeys the general law.

2. The apparent reason for this change in polymorphic modifications as temperature rises is the increase in the energy of electrons in the outer shells of the ions composing the metal lattices. This leads to the disruption of the directional bonds and a transition from covalent structures to metallic, and from cavalent—metallic composite structures to the typical cubic structures of metals, as well as to a reduction in the ellipticity of ions with outer s- and d-electrons and to a transition from hexagonal close-packed to face centered cubic structure as the shape of the outer electron clouds approaches spherical.

3. At the highest temperature, close to the melting points, the ions in the crystal structures of all the metals in Groups Ia, IIa, IIIb, and IVb, including the lanthanides and actinides, lose all the outer valence electrons, and, as a result of the consequent uncovering of the orthogonal p^6-shells, body-centered cubic structures are realized.

Literature Cited

1. N. N. Sirota and E. M. Gololobov, Dokl. Akad. Nauk SSSR, 138(1): 162(1961); 143(1):156(1962); 144(2):398(1962).
2. E. M. Savitskii, Effect of Temperature on the Mechanical Properties of Metals and Alloys (Izd. Akad. Nauk SSSR, 1957).
3. L. Pauling, Phys. Rev., 54:889 (1938); Proc. Roy. Soc., A 196:343 (1949); Acta. Cryst., 4:138 (1951).
4. W. Hume-Rothery, Atomic Theory for Metallurgists [Russian translation] (Metallurgizdat, 1955).
5. W. Delinger, Theoretical Physical Metallurgy [Russian translation] (Metallurgizdat, 1960).
6. F. H. Spedding, J. J. Hanak, and A. H. Daane, J. Less-Common Metals, 3(2):110 (1961).
7. W. B. Pearson, Handbook of Lattice Spacing and Structures of Metals and Alloys (Pergamon Press, London, 1958).
8. V. K. Girgorovich, Periodic System of D. I. Mendeleev, Structure and Properties of the Elements. Handbook, "Physical Metallurgy and Heat Treatment" Vol. I (Metallurgizdat, 1961).
9. V. K. Grigorovich, Transactions of the A. A. Baikov Institute of Metallurgy, No. XIV (Moscow, Izd. Akad. Nauk SSSR, 1963).

GROWTH OF A SINGLE CRYSTAL FROM THE SOLID PHASE

DURING A POLYMORPHIC TRANSFORMATION OF n-DICHLOROBENZENE

A. I. Kitaigorodskii, Yu. V. Mnyukh, and Yu. G. Asadov

The present investigation is the first of a planned series devoted to the mechanism of polymorphic transformations in organic molecular crystals. The authors have chosen as their first object of study n-dichlorobenzene. This substance has already been repeatedly studied by various scientists (See [2, 3]). It has been established that there is a polymorphic transformation of the low-temperature α phase into the high-temperature β phase at 30.8°C. The structures of these crystalline phases have been determined [4, 5], with the crystals grown in each case from a melt at the equilibrium temperature for the modification under examination.

The low-temperature α phase has a monoclinic structure with unit cell parameters

$$a = 14.80 \text{ Å}, \qquad \angle \beta = 113°$$
$$b = 5.78 \text{ Å}, \qquad Z = 2,$$
$$c = 3.99 \text{ Å},$$

and the high-temperature β phase has a triclinic structure with unit cell parameters

$$a = 7.32 \text{ Å}, \qquad \angle \alpha = 93°10',$$
$$b = 5.95 \text{ Å}, \qquad \angle \beta = 113°35', \; Z = 1,$$
$$c = 3.98 \text{ Å} \qquad \angle \gamma = 93°30'.$$

Visual observations [3, 6, 7, 9] suggest that the polymorphic transformation takes place by an unknown single-crystal-to-single-crystal process. The rates of transformation were measured in [3, 7, 9].

The first part of the investigation described here consisted of observing the $\alpha \rightleftharpoons \beta$ transformation processes under the optical microscope, furnished with a heating stand, a polarization attachment, and an arrangement for taking microphotographs.

Since solid n-dichlorobenzene is highly volatile, special attention was paid to the choice of a liquid inert medium in which crystals of this substance could be preserved for an extended period. This requirement proved to be fully satisfied by glycerine. In examination under the microscope, the crystals were placed in an open glycerine-filled cuvette with low walls, the bottom of which served as object glass for the microscope. Some n-dichlorobezene of the "pure" type was submitted to additional purification by vacuum distillation. In contrast to many previous investigations, great attention was paid in the present work to the perfection of the crystals studied.

Observations showed that the $\alpha \rightleftharpoons \beta$ transformation always takes place at a higher temperature than that of phase equilibrium, 30.8°C. In general, the following rule holds: The purer the material and the more perfect the crystal, the higher is the $\alpha \rightleftharpoons \beta$ transformation temperature. Perfect crystals specially produced by sublimation in most cases failed to pass into the β-phase at all, but melted in the α-form at 52.7°C (the β-form melts at 53.2°C).

The $\alpha \rightleftharpoons \beta$ phase transformation always appears in the form of a well-defined phase-separation boundary, especially in polarized light. All cases may be divided into two groups, "single-centered" and "multicentered" transformations. It appears impossible to predict in advance which type of transformation will be observed in a given crystal (if the original crystal is fairly perfect). If no growth center of the new phase is formed, then, as already indicated, the transformation does not take place. If just one growth center is formed, there is a single-crystal-to-crystal transformation. Upon the development of several growth centers, a phase transformation takes place with the formation of a system of crystallites, the orientation of which will concern us shortly.

Single-crystal-to-single-crystal transformations, which can only arise for fairly perfect crystals, are shown in two series of microphotographs (Figs. 1 and 2). The photographs shown indicate that the single-crystal-to-single-crystal polymorphic transformation is nothing but the growth of a faceted single crystal from a solid single-crystal medium of nonequilibrium phase. The growth of single crystals from the solid phase is very strongly reminiscent of the growth of single crystals from liquid and gaseous media.

Figure 1 shows two out of a series of successive microphotographs of the $\alpha \rightleftharpoons \beta$ transformation in a single crystal of acicular form obtained by sublimation. The arrows indicate the direction of growth. This direction could be changed at will by raising or lowering the temperature. One can see the perfect form of the phase-separation boundary; in the growing single crystal this always has convex form, indicating its tendency toward minimum surface free energy. We note that the facing of the phase-separation boundary is the more perfect, the slower the growth of the equilibrium phase crystal. The rate of growth depends markedly on the peculiarities of the specific crystal and its history. In particular, this rate may be reduced and even halted by repeated changes in the direction of the phase transformation. In this way we can fix the phase-separation boundary in a definite position and make it insensitive to any temperature variations. We may suppose that this phenomenon is due to an accumulation of stresses near the phase-separation boundary. This is also indicated by the iridescent halos clearly visible in Fig. 1. It was noted that the halo was always directed from the phase-separation boundary to the growing-crystal side. "Relaxation" of the crystal for several days leads to a partial resorption of these stresses, as a result of which the possibility of displacing the phase-separation boundary into a new position (by changing the temperature) is partly restored. The following observation is also of interest: At any moment the α- and β-phases may be disconnected mechanically and the phase-transformation process stopped, since the crystal breaks easily along the surface of separation. Moreover, the fracture preserves the facing of the surface of separation. Figure 2 shows the growth process of a well-faced β-single crystal inside an α-single crystal, from its genesis to the end of the phase transformation. The seed was generated at 48°C and grew extremely rapidly; the intervals between photographs were a few tens of seconds.

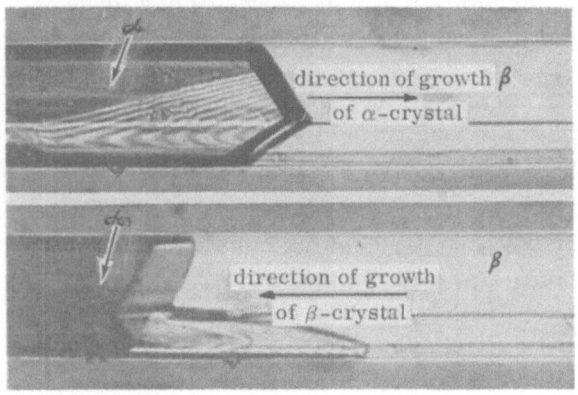

Fig. 1. $\alpha \rightleftharpoons \beta$ transformation in a single crystal of acicular form.

The next experimental problem was to determine the mutual orientation of the α- and β-lattices. We had to answer the questions: 1) Is there a single (rigid) connection between the orientations of the α- and β-crystal lattices? 2) If there is, what are the mutual orientations? 3) If there is not, then is there a discrete set of possible mutual orientations, or have they an arbitrary character?

To answer these questions, we used Laue X-ray photographs of one and the same crystal before and after phase transformation, maintaining the external orientation of the crystal. For this, a well-formed α-phase single crystal (growth in the usual way from solution in ethyl alcohol) was placed in a small, thin-walled, glycerine-filled flat cuvette, which was installed on a goniometer head (Fig. 3). Careful adjustment was made by means of Laue photographs, as it was important to try and get exact coincidence

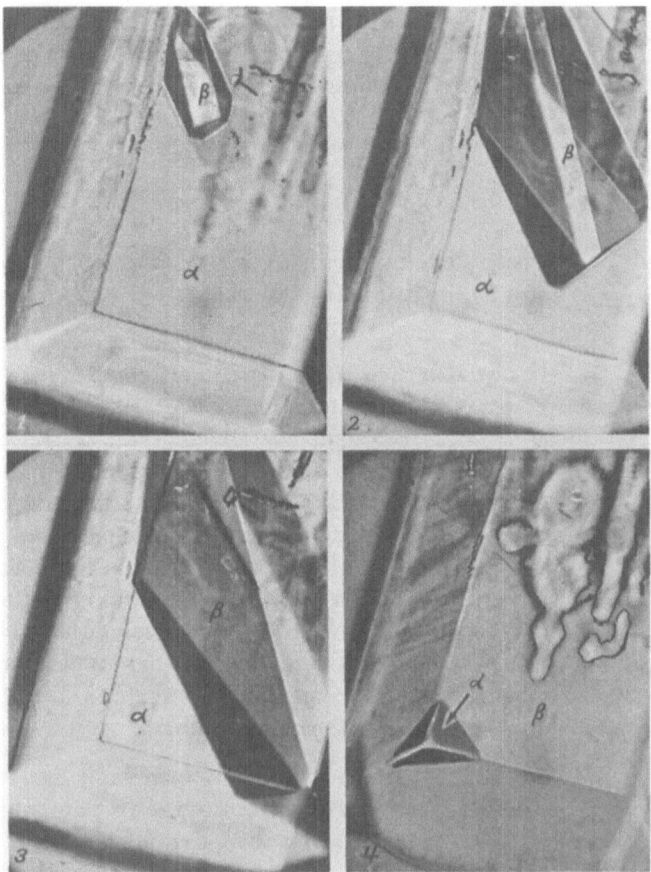

Fig. 2. Successive stages in the growth of a faceted single crystal of the β-phase inside a nonequilibrium α-phase crystal.

between the initial orientations of all the crystals examined. The X-ray camera was placed in a specially constructed air thermostat, in which the raised temperature needed for the $\alpha \rightarrow \beta$ transformation was maintained, and then a Laue photograph of the β-phase was taken.

This X-ray investigation presented a number of experimental difficulties which had to be overcome. First of all, there was the inconvenience of photographing crystals in a liquid-filled cuvette. The soft, easily deformed n-dichlorobenzene crystal had to be fixed inside the cuvette without producing stresses and strains, but in such a way as to ensure precise maintenance of the external orientation before and after the phase transformation. The adjustment of the crystals had to be effected solely by means of the Laue photographs, without the aid of the optical goniometer. Another feature of the experimental difficulty was due to the nature of the subject examined. We have already pointed out that the phase transformation in many cases could not be induced in the crystal. But even in cases where it took place, it was certainly not always single-centered. Moreover, X-ray pictures of an extremely large number of crystals were required.

Finally, however, we were able to establish quite reliably from the X-ray results that there was a l a r g e n u m b e r   o f   l a t t i c e   o r i e n t a t i o n s   o f   t h e   n e w   p h a s e   r e l a t i v e   t o   t h e   o l d . There are good grounds for asserting that the β-crystals grow with completely arbitrary orientation relative to the lattice of the α crystals, since the number of orientations obtained was very large.

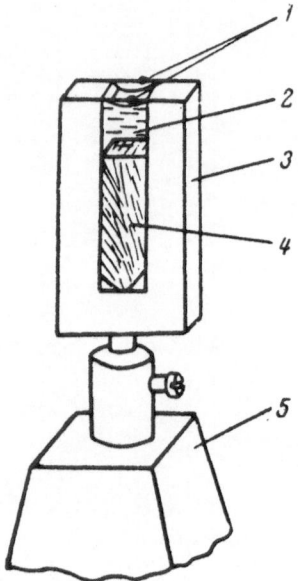

Fig. 3. System for X-ray photography of n-dichlorobenzene crystals immersed in an inert liquid; 1) thin organic film; 2) glycerine; 3) organic glass frame; 4) crystal; 5) gonimeter head.

An experiment which proved to be extremely convincing involved one of the single crystals of moderate size ($1 \times 3 \times 8$ mm), in which we were able to induce ten transformations:

$$\alpha_1 \ (\text{original}) \rightarrow \beta_1 \rightarrow \alpha_2 \rightarrow \beta_2 \rightarrow \alpha_3 \rightarrow \beta_3 \rightarrow \alpha_4 \rightarrow \beta_4 \rightarrow \alpha_5 \rightarrow \beta_5 \rightarrow \alpha_6.$$
$$\downarrow$$
$$\beta_5$$

These phase transformations are described by the Lane photographs of Fig. 4 and by the data in the tables.

Among the β phase shown in the table, there are no two coinciding in crystal lattice orientation. This conclusion is confirmed by X-ray photographs of many other crystals. We also see from the table that in the $\alpha \rightarrow \beta \rightarrow \alpha \rightarrow \ldots$ sequence one and the same crystal may give both single-centered and multicentered transformations, in the second case the single crystal transforming into a system of differently oriented crystals of the other phase. A perfect single crystal may again grow subsequently from this disorganized crystal. In fact, the return to the perfect α-phase crystal takes place not only after single-centered but also after multicentered transformation. Visual as well as X-ray observations show that for a multicentered $\alpha \rightleftharpoons \beta$-transformation the crystal consists of a series of distinct blocks. On subsequent single-centered transformation, the block boundaries vanish and a monolithic crystal again develops.

It also seemed extremely interesting to us that the orientation of the α-crystals almost always returned to its original state with a precision of 1°. This naturally raises the question as to where the information on the orientation of the α-lattice is stored. The phenomenon of the regeneration of the α-lattice was observed in crystals grown from solution in the α-form. This gave grounds for supposing that the external faces constituted the "memory" element. Preliminary experiments show, however, that the external facing of the crystal cannot contain the remembered information: A series of Laue photographs analogous to those shown in the table but obtained from a crystal grown from solution in the β-form, again showed a variety of orientations of the β-crystals and coincident orientations of the α-crystals.

The X-ray part of the present investigation has led us to two results in clear contradiction to one another. On the one hand, in the large number of $\alpha \rightarrow \beta$ transformations studied we never found two reproducible cases of mutual orientation of the α and β crystals. This would suggest a diffusion-disordered transformation. On the other hand, transformation of the $\alpha \rightarrow \beta \rightarrow \alpha \rightarrow \ldots$ type in the majority of cases gave α-crystals in one and the same orientation. This means that the course of the $\alpha \rightarrow \beta$ transformation is, as it were, "remembered" by the β-crystal and reproduced on its return into the α-form, and suggests a regular rearrangement of the lattice. Investigations continuing at the present time should elucidate the cause of this contradiction.

Before carrying out this investigation, the authors considered the most probable of the hypothetical possibilities indicated to be a "rigid" connection between the orientations of the α and β lattices. For this supposition there were, as it appeared, important factual bases, namely, the similarity of the molecular packing in the α and β structures. Witness to this is the similarity between the unit cell parameters given above (the difference lies mainly in the transformation from the one-layer packing of the β form to the two layer of the α, which doubles the a spacing). Post factum, however, we can see a direct logical connection between the results obtained and the well-known case of the single-crystal-to-polycrystal polymorphic transformation, when the single crystal transforms into a system of a large number of differently oriented crystallites. It is

X-Ray Characteristics of Successive $\alpha \rightarrow \beta \rightarrow \alpha \rightarrow \ldots$ Transformations in One of the
n–Dichlorobenzene Single Crystals

Crystal form	Transformation	Lattice orientation	Form of Laue spots
α_1	Original crystal (grown from solution in ethyl alcohol)	Symmetric Laue photograph c axis vertical	Form typical of perfect crystal
β_1	Single-centered	Not determined; c axis not vertical	Form typical of perfect crystal
α_2	Single-centered	Coincides with α_1	Spots split. Mutual disorientation of two lattices $0.7 \pm 0.2°$
β_2	Multicentered	—	Some spots regular form, rest split or blurred
α_3	Single-centered	Coincides with α_1	Spots split as in α_2
β_3	Single-centered	Differs greatly from β_1	Form typical of perfect crystal
α_4	Single-centered	Coincides with α_1	Irregular form of spots
β_4	Single-centered	Like β_3 but differing by $\sim 5°$	Spots slightly split
α_5	Single-centered	Coincides with α_1	Form typical of perfect crystal
β_5	Multicentered	Differs from β_2 and partly noncoincident with other β lattices	Laue photograph consists of a large number of small spots
$\beta_5^{!}$[*]	Mainly single-centered	Differs from $\beta_1 - \beta_4$. Contained as a part in β_5	Some spots split; weak blurred spots also present
α_6	Single-centered	Differs from all the previous	Form typical of very perfect crystals.

[*] This Laue photograph was taken for another part of the crystal.

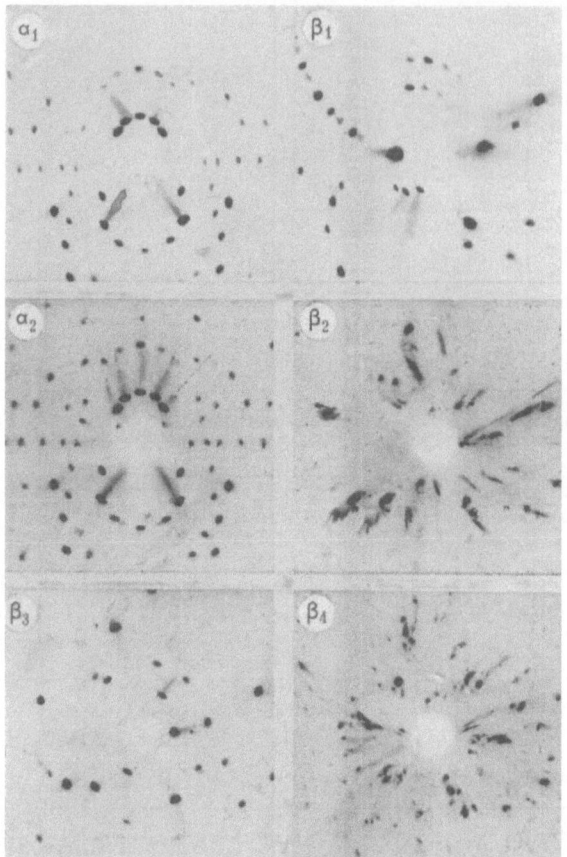

Fig. 4. Laue photographs illustrating the polymorphic
transformations in an n-dichlorobenzene crystal.

extremely probable that the difference between single-crystal-to-single-crystal and single-crystal-to-polycrystal phase transformations may reduce simply to the conditions determining the number of crystallization
centers generated.

Literature Cited

1. A. I. Kitaigorodskii, Yu. G. Asadov, and Yu. V. Mnyukh, Dokl. Akad. Nauk SSSR, 148:1065 (1963).
2. M. F. Vauks, Zh. Eksp. i Teor. Fiz., 7:270 (1937).
3. V. I. Danilov and D. E. Ovsienko, Dokl. Akad. Nauk SSSR, 73:1169 (1950).
4. J. Housty and J. Clastre, Acta Cryst., 10:695 (1957).
5. E. Frasson, Acta Cryst., 12:126 (1959).
6. M. F. Vuks, Dokl. Akad. Nauk SSSR, 1(10):69 (1936).
7. B. Lemanceau and C. Clement, Compt. Rend. Acad. Sci., 248:3157 (1959).
8. A. I. Bykhovskii, L. N. Larikov, and D. E. Ovsienko, Kristallografiya, 6:248 (1961).
9. G.B. Ravich and O. F. Bogush, Izv. Sektora Fiz. Khim Analiza, 23:309 (1953).

PHASE TRANSFORMATIONS IN NAPHTHALENE CRYSTALS

V. L. Broude, M. S. Soskin, and A. K. Tomashchik

It was shown in [1, 2] that, with the help of spectral methods, the changes taking place in crystals during deformation could be monitored, and, quite apart from changes in lattice spacing, the changes in the orientation of individual molecules in the crystal could be studied. All effects associated with the deformation of crystals are revealed in a more pure form at low temperatures, when the thermal vibrations of the lattice are considerably weakened and the absorption bands discrete.

For investigation we chose a naphthalene crystal, as the best known of all molecular crystals. At 20°K, the absorption spectrum of the unstressed crystal of naphthalene consists of a large number of isolated bands [3] (Fig. 1). The band of the purely electronic transition of the naphthalene molecule is split in the spectrum of the crystal, forming an exciton doublet of absorption bands, the components of which are polarized along the axis of the crystal (A_1 band with frequency 31,476 cm^{-1} and B_1 band with frequency 31,623 cm^{-1}). The exciton doublet has a clearly visible vibrational echo (A_2 band with frequency 32,231 cm^{-1} and B_2 band with frequency 32,261 cm^{-1}). The vibrational frequency equals 715 cm^{-1}. These two doublets, together with additional bands hidden among other strong bands of the spectrum, form a series of exciton bands. The remaining weakly polarized bands arise for local excitations of the crystal and form the so-called molecular series [4]. The intensity ratio of the two components of the molecular absorption band (polarization ratio) is determined by the orientation of the dipole moment of the electron-vibrational transition in the crystal.

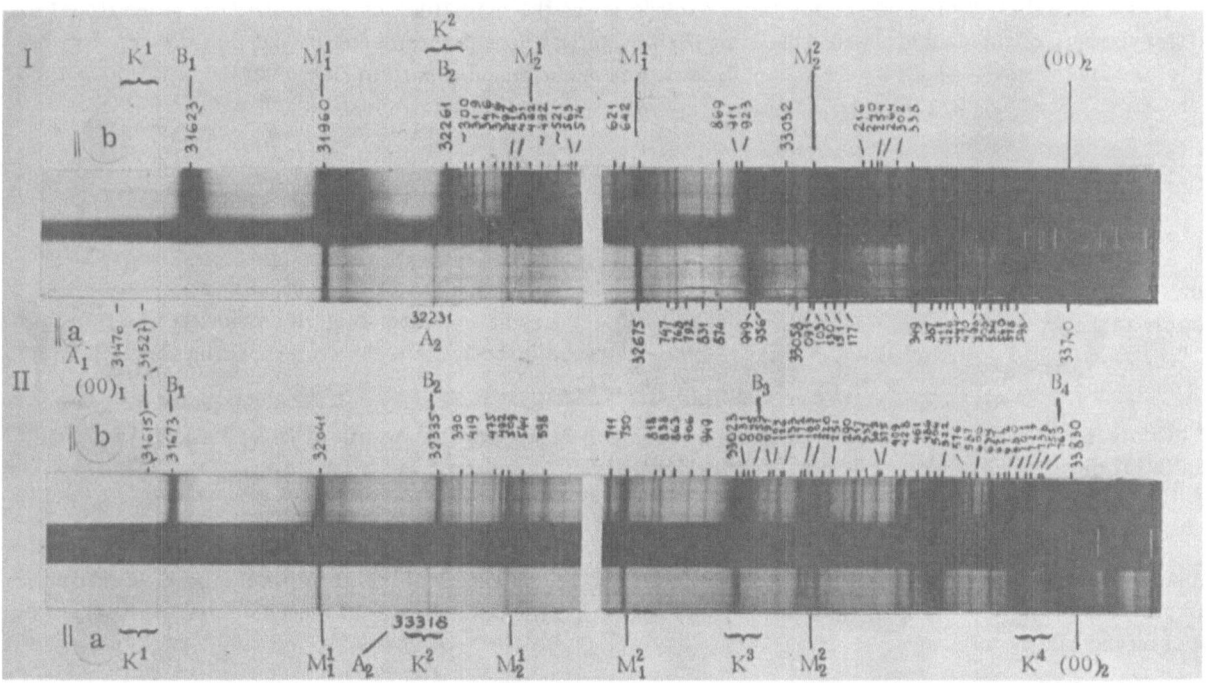

Fig. 1. Absorption spectra of naphthalene crystals at 20°K; I) spectrum of free crystal; II) spectrum of a crystal about 0.1 μ thick grown in the curvette.

The naphthalene samples studied were grown from the melt in a special quartz cuvette. They constituted single-crystal films of various thicknesses (from 0.5 to 1.5 μ), rigidly fixed by an optical contact to the cuvette walls. The cuvette was fixed to a rod and cooled by liquid hydrogen or nitrogen in a special optical cryostat [5]. With the aid of a microprojector, various parts of the crystal could be projected on to the spectrograph slit. Upon cooling of the cuvette, strain developed in the crystal, since the temperature coefficient of expansion of the quartz was approximately 1000 times smaller than that of the crystal. The stresses associated with the strain in the sample could be judged by comparing the spectra of unstrained crystals with those of equally thick crystals fixed in the cuvette. In the latter case, all the bands in the absorption spectrum shifted, and in some cases changed their form and half-width [1, 2].

First of all, we must elucidate the character of the stresses arising in such samples. Since the ratio of the longitudinal dimensions of the crystals to its thickness is approximately 10^4, the stresses must be homogeneous. A certain inhomogeneity arises near the ends of the crystal at distances of the order of its thickness, and this may be excluded from consideration. Thus the strain may be described as a plane homogeneous extension of the sample. To estimate the extent of the strains, we used the connection between the intermolecular distance and the exciton splitting, which for the naphthalene crystal is determined by the octupole−octupole term in the expansion of the interaction energy between molecules in terms of multipoles, and hence depends on the distance as $1/r^7$ [12].

The extent of the splitting may be determined primarily from the data of [7] from the value of the expansion coefficient of naphthalene, by assuming that, on lowering the temperature, optical contact is maintained between sample and cuvette. In fact, if in the unstressed crystal the splitting for the A_1-B_1 band equals 147 cm^{-1}, then in the crystal extended by some 5% (which is realized on cooling the sample to 20°K) it should be $147 \cdot 1/(1.05)^7 \simeq 100$ cm^{-1}.

It is known that within one and the same exciton series the magnitude of the splitting is proportional to the total strength of the doublet band oscillators [6]. Quantitative measurements showed that the ratio of the sums of the intensities of bands $A_1 + B_1 / A_2 + B_2$ for a free and stressed superthin (less than 0.1 μ thick) sample are the same and equals 3/1. Hence we have

$$\frac{(A_1, B_1)_{\text{free}}}{(A_1, B_1)_{\text{thin}}} = \frac{(A_2, B_2)_{\text{free}}}{(A_2, B_2)_{\text{thin}}} ,$$

where $(A_1, B_1)_{\text{free}}$ and $(A_2, B_2)_{\text{free}}$ are the exciton splittings for the doublet of bands A_1, B_1 and A_2, B_2 in the spectrum of the free crystal, and $(A_1, B_1)_{\text{thin}}$ and $(A_2, B_2)_{\text{thin}}$ are the corresponding values for the the thin sample. $(A_1, B_1)_{\text{free}}$, $(A_2, B_2)_{\text{free}}$, and $(A_2, B_2)_{\text{thin}}$ can be determined directly from the corresponding spectra. Then, using the proportionality, we obtain for $(A_1, B_1)_{\text{thin}}$ the value 95 cm^{-1}.

Comparison of the low-temperature spectra of naphthalene with the spectra of its vapor and solutions given in [4] gave for $(A_1, B_1)_{\text{thin}}$ 86 cm^{-1}. Thus different estimates predict for $(A_1, B_1)_{\text{thin}}$ the value 95 ± 10 cm^{-1}.

Direct experimental determination of the splitting for the thin crystal is impossible, since for thicknesses of approximately 0.1 μ the A_1 band is not visible, owing to the weakness of the absorption, and a thicker crystal (as in the case of the free sample) must not be used, since the stresses arising and the character of the spectrum depend on the sample thickness.

A pile of 20 cuvettes with thin naphthalene crystals of approximately the same thickness was therefore assembled, and their absorption spectra photographed in unpolarized light. The total thickness of the napthalene in this pole was 2 to 2.5 μ, and the A_1 band for this crystal thickness was clearly visible (Fig. 2). The splitting determined from the spectrum was 95 ± 20 cm^{-1}. The imprecision is due mainly to the fact that the crystals making up the pile were not strictly of the same thickness.

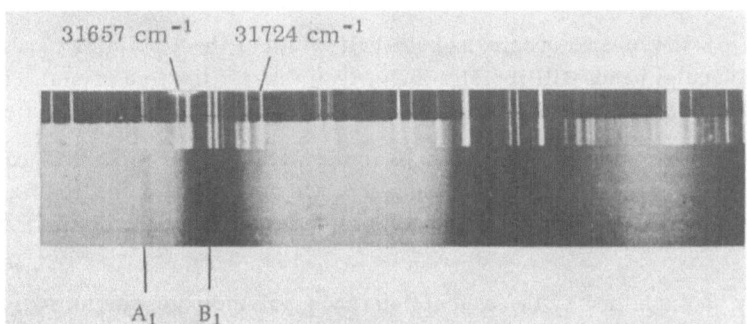

Fig. 2. Absorption spectrum of a pile of 20 thin naphthalene
crystals.

Thus the agreement between the three theoretical estimates and the experimental value of exciton splitting for the A_1, B_1 doublet enables us to conclude that it is in fact approximately 90 cm^{-1}.

These facts thus lead us to the conclusion that, upon cooling of the crystal in the cuvette, there is in fact strain in the sample having the nature of an elongation determined by the difference between the coefficients of expansion of the crystal and the quartz.

We made qualitative and quantitative studies of the spectra of a large number of naphthalene crystals in the free and stressed states. We measured the position of the B_1 band and one of the molecular bands (in the spectrum of the free crystal its frequency equalled 31,960 cm^{-1}). The results are shown in Fig. 3 (the figure shows the b component of the M band, since the splitting of this band is observed in the stressed crystal [1]; an analogous relationship holds for the a component). As we see, the experimental points form two branches between which there is no continuous transition, which may indicate the spasmodic nature of the changes taking place in the crystal as its thickness varies smoothly. In the very thin samples obtained in the cuvette, we may expect changes in certain parameters characteristic of massive crystals. This also leads to a variation of the stresses arising in the sample with its thickness. Over a certain range of thicknesses this relationship is linear, in agreement with earlier data on the deformation of naphthalene crystals under certain other conditions [7].

A comparison of the spectra of stressed crystals of different thickness showed that, as the sample thickness changed monotonically, there was a nonmonotonic change in the half-width of individual spectral bands. In the spectra of crystals between 0.3 and 1μ thick the half-width of both molecular and exciton absorption bands rises above that found in the spectra of free crystals. On further reduction of the sample thickness, however (to thick-

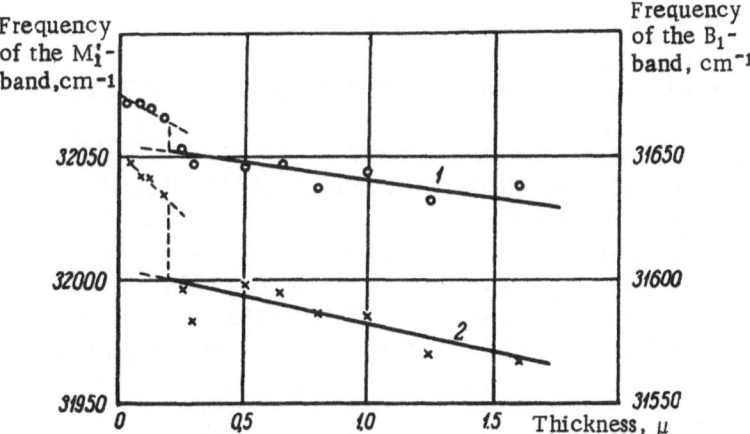

Fig. 3. Variation of the position of the absorption bands of
the rigidly fixed naphthalene crystals with sample thickness:
1) for the exciton B band; 2) for the molecular M band.

nesses of the order of 0.1 μ), the reverse occurs, and the half-width of the absorption bands becomes smaller. The half-width of the molecular bands still remains larger than that for the free crystal. The half-width of the exciton B_1 band in the spectrum of the most stressed crystal is even smaller than for the free crystal [8].

We notice the fact that the nonmonotonic changes in the widths of the bands and the spasmodic changes in their positions take place for the same sample thicknesses. On deformation, the polarization ratio for the molecular bands changes, which may indicate a change in the orientation of the dipole moment of the electron-vibrational transition.

All these facts may be explained by the assumption that a polymorphic transformation takes place in the naphthalene crystals under the influence of stresses increasing with diminishing sample thickness.

As we know, the relation between the position of the absorption edge and pressure has been studied in a number of investigations [9, 10]. Moreover, we know that polymorphic transformations in the thoroughly studied alkali halide and silver halide crystals is accompanied by a spasmodic displacement of the absorption edge.

By analyzing the results obtained, we can make certain conjectures regarding the transformation observed. This must be related to the diffusionless type, since on deformation the single crystal nature of the sample is preserved. This is indicated by the external similarity of the spectra from the free and thin samples (see Fig. 1); the absorption bands in both cases are narrow, with sharply expressed polarization. The sharp polarization of the bands in the exciton doublet, retained in spectra from crystals of all thicknesses, moreover, confirms that free excitons with finite diffusion length can be excited in such samples. The reduction in the half-width of the exciton bands in the thin crystal noted above indicates a change in exciton—phonon interaction [1]. At the same time, by comparing the half-widths of the molecular bands sensitive to a local change in the lattice, we find that the crystal formed under experimental conditions is less perfect than the original.

In conclusion, we must note that the very fact of having detected a polymorphic transformation in the naphthalene crystal from the form of its absorption spectrum, together with other results regarding the nature of the changes taking place in the crystal on cooling in the cuvette, clearly illustrates the rich possibilities opened by studying the exciton characteristics of crystal spectra.

Literature Cited

1. V. L. Broude, O. S. Pakhomova, and A. F. Prikhot'ko, Zh. Opt. i Spektr., 2:318 (1957).
2. V. L. Broude, V. V. Eremenko, V. S. Medvedev, O. S. Pakhomova, and A. F. Prikhot'ko, Ukr. Fiz. Zh., 3:232 (1958).
3. A. F. Prikhot'ko, Zh. Eksp. i Teor. Fiz., 19:383 (1949).
4. V. L. Broude, Zh. Opt. i. Spectr., in press.
5. V. P. Babenko, V. L. Broude, V. S. Medvedev, and A. F. Prikhot'ko, Pribory i Tekhn. Eksperim., (1):115 (1959).
6. A. S. Davydov, Tr. Inst. Fiz. Akad. Nauk, No. 1, 1951.
7. A. E. Uinderhorn and H. G. Drikamer, Phys. Chem. Solids, 9:330 (1958).
8. M. S. Soskin, Ukr. Fiz. Zh., 5:707 (1960).
9. Y. Toyozawa, Progr. Theor. Phys., 20: 53, 1958.
10. T. E. Slykhouse and H. G. Drikamer, Phys. Chem. Solids, 7:207 (1958).
11. R. A. Eppler and H. G. Drikamer, Phys. Chem. Solids, 6:180 (1958).
12. D. P. Craig, L. E. Lyons, S. H. Walmsley, and J. R. Walsh, Proc. Chem. Soc., p. 389 (1959).

ROLE OF STRUCTURAL IMPURITIES IN PHASE
TRANSFORMATIONS IN SOLIDS

A. I. Bykhovskii

Analysis of a series of experimental data on the kinetics of phase transformations leads to the conclusion that centers of the new phase are generated at certain "prepared sites." Recently the idea has become widespread that various structural impurities play the part of such "prepared sites" for phase transformations in crystalline media [1]. In the case of the crystallization of liquids, undissolved particles constitute the structural impurities. For phase transformations in solids, besides particles of extraneous phases, a similar role is also played by dislocations, grain boundaries, and other nonequilibrium deviations from structural regularity.

It was shown in [2] that, on the frequency curves of the radii of pearlite columns, instead of a straight line parallel to the axis of abscissas characterizing a homogeneous formation of nuclei, there was a sharp maximum indicating the existence of "prepared sites" during the transformation of austenite into pearlite [3]. The existence of stable fixed points for the generation of martensite crystals in the original β_1 phase of copper—tin and copper—aluminum alloys is indicated by microstructural studies of transformations in these alloys [4].

The transformation of white tin into gray begins at shear or puncture points [5]; the yellow crystals of mercury iodide, which exist at room temperature in a metastable state, rapidly transform into crystals of the stable red form under the action of a puncture [1].

Plastic deformation also increases the generation rate of centers in the low-temperature $\beta \rightarrow \alpha$ transformation of uranium and its alloys [6]. The accelerating function of plastic deformation is due to the increase in the free energy of the deformed phase and hence to the increase in the difference of free energy between the old and new phases, i.e., in the thermodynamic "stimulus" of the transformation, and also to the increase in the diffusion mobility of the atoms resulting from the plastic deformation. For other transformations in uranium [6] and for a number of martensite transformations, the accelerating role of grain boundaries is well known.

There is also quite a lot of information in the literature as to the accelerating role of second-phase particles in the phase transformations of solids. The effect of insoluble impurities artifically introduced into the original material on the austenite decomposition process was demonstrated in [7]. The author carburized steel samples containing 0.5% aluminum from a gaseous medium containing oxygen and from another oxygen-free medium. As he worked with the oxygen-containing medium, Al_2O_3 particles developed, and these had a strong nucleating effect on the formation of pearlite centers. In the oxygen-free medium this did not happen.

In the $\beta \rightarrow \alpha$ transformation of tin, it was noticed that the presence of second-phase particles in the samples accelerated the formation of centers of gray tin [8—10]. It was shown in [11] that, in tin–germanium alloys with germanium concentrations considerably exceeding the solubility limit in metallic tin, at various low temperatures the time preceding the spontaneous appearance of α–Sn was sharply curtailed. However, simply rubbing Ge crystallites and a number of other materials into white tin [12] did not markedly accelerate its transformation into gray. Evidently the difference between the artificially introduced insoluble impurity and the second-phase dispersed particles noncoherent with the matrix is that the latter particles are surrounded by an amorphized layer and create large microstresses in the matrix. Some very interesting results on attaining the $\beta \rightarrow \alpha$ transformation of tin with the aid of isomorphous seeds were obtained in an investigation of Goryunova [12]. It was shown that, together with gray tin itself, the role of "seed" in the $\beta \rightarrow \alpha$ Sn transformation could also be played by the semiconducting compounds CdTe and InSb, which like α—Sn have the diamond-type lattice and parameters similar to those of α—Sn. This result is analogous to the accelerating effect of active impurities on the crystallization of liquids, earlier unknown for artificially introduced inclusions in solid transformations. In [10] particles of CdTe, InSb, Si, and Ge were uniformly introduced into the surface layer of tin by baking-in; these proved to be sites for the preferential development of α-Sn centers. During the generation of α-Sn centers in Sn + CdTe and Sn + Si samples, at certain supercoolings all or nearly all the artificially introduced "prepared sites" were exhausted, which was manifested by the nonmonotonic behavior of the surface diffusion coefficient of mercury on tin at low temperatures [9, 10].

It is possible that insoluble impurities play an important part in a number of other cases of phase transformations. Thus it was shown in [13, 14] that for small superheatings in the case of the $\alpha \rightarrow \gamma$ Fe and $\alpha \rightarrow \beta$ U transformations the original single crystals of the low-temperature phase transformed into polycrystals, but after cooling were restored. This property is easily seen in samples with nitrogen and carbon impurities [13, 14], in which the greater the superheating, the greater the concentration of the small N and C impurities. The authors explained this regeneration by the retention of low-temperature phase seeds during superheating.

If we start from the concept of the existence of "prepared sites" for solid-state phase transformations, then it is natural to suppose that there is a certain distribution of these "prepared sites" with respect to activity. This activity must be determined by the physicochemical nature of the active part and by geometrical factors (shape, size). It is natural that for large supercoolings (or superheatings) the centers of the new phase will be generated at less active parts, but the growth of these centers may be seen at lower supercoolings. A method similar to the Tammanov method of "revealing" centers during the crystallization of liquids was used, in particular, for studying polymorphic transformations of sulfur [15] and plutonium [16]. The authors of [15], in studying the transformation of monoclinic sulfur into rhombic (transformation point 95.6°C) held it for 20 min at -20°C and then dilatometrically determined the kinetics of the transformation at 32°C. It was noted in [16] that if, after finishing the transformation of β-Pu into α-Pu at a given temperature, the samples were quenched in a bath at -21°C and then heated to a temperature in the range 26 to 75°C, then an additional quantity of the α phase was formed, increasing as the holding temperature was raised. The possible role of impurities in the transformation is indicated by the fact that the effect in question occurs more strongly for less pure specimens.

Effect of the Thermal History of the Samples

In view of the fact that the thermal history of samples may occasion a redistribution of components during phase transformations in alloys, let us consider the effect of the previous heat treatment of the samples and recrystallization on the kinetics of polymorphic transformations. Such an effect takes place in many cases. Thus it was shown in [17-22] that the rate of the $\beta \rightarrow \alpha$ transformation of tin increases if the transformation has occurred previously in the material used. In such cases we speak of the "remembrance" or "memory" of the sample, of the effect of thermal history or thermal biography of the sample on subsequent transformation, and so forth.

It was shown in a number of papers [23-25] that in the case of the crystallization of liquids the pre-history of the samples is felt only in the generation of crystallization centers on the surface of activating insoluble impurities. It was shown in [25] that crystallites of a given phase situated in micropores of insoluble impurities, in molecular contact with their walls, may be preserved in these micropores at temperatures considerably higher than the melting point. Analogous concepts may also be applied to solid-state transformations. Thus it was shown in [26] that, in certain conditions, crystals of a phase metastable in bulk will be stable in the microvacancies of inclusions, and may constitute reorientation seeds on recrystallization or on nuclear precipitation of a new phase. In these cases the presence of second-phase inclusions may facilitate the reverse as well as the forward transformation. Support for this explanation of the "memory" effect in tin transformations comes from the fact that neither the forward nor the reverse transformation goes to completion, some 1 to 2% of the original phase always remaining untransformed, as indicated by density measurements [21, 27-29]. Thus from the data of [29], for previously fused white tin we have at room temperature $\rho_{\beta\text{-Sn}} = 7.285$ g/cm^3. If gray tin is held at 30°C for 10 to 12 h, then its density is lower: $\rho = 7.245$ g/cm^3, (i.e., 2.6% α-Sn is not converted); this does not change on further holding at this temperature, but after 60 h at 100°C (or 27 h at 120°C) ρ returns to 7.267 g/cm^3 (1.2% α-Sn is not converted).

If we suppose that this action of extraneous inclusions really does take place in solid-state transformations, then we may also expect the effects of activation and deactivation of the impurities, as in the analogous effects during crystallization of a liquid on insoluble activated impurities. Such effects are indeed observed. Thus it was shown in [30] that the kinetics of the polymorphic transformation of HgI_2 depend on the thermal history of the sample. The rate of transformation of all samples increased with increasing number of previous transformations. This increase is especially noticeable after the first few transformations. The following data were given in [30]: The maximum overall rate of transformation of red mercury iodide into yellow (other conditions being equal) was five times greater at the sixth transformation than at the first, and seven times greater at the tenth. In succeeding experiments the activity changed less and less and finally reached a constant value. In another series of experiments the effect of the temperature and holding time preceding the transformation on the transformation rate was noted. The transformation of yellow HgI_2 into red at 120.2 and 120.6°C was studied (transformation point between 126 and 127°C). Previously, maximum activity was imparted to the samples by a large number of recrystallizations. It was found that, on increasing the holding time of the samples at 137°C (in the yellow form) from 1 to 5 and from 1 to 14 h respectively, the maximum overall transformation rate fell substantially, and the time to reach this maximum rate as well as the time for the whole transformation correspondingly increased.

The experiments mentioned are similar to those of Danilov et al. on the crystallization of liquids. The effect of the thermal history of samples on the kinetics of transformations would seem to indicate the generation of new-phase centers on structural impurities. From this point of view the first series of experiments with HgI_2 may be interpreted as activation of these impurities with increasing number of recrystallizations, until all possible "prepared sites" for the supercooling in question had been used (maximum activity of the samples), and the second series of experiments as the deactivation of these "prepared sites" at a higher temperature.

It was shown in [31] that cyclical heat treatment with transitions through the equilibrium temperature of the two phases in tin affects the character of the transformation of white tin into gray. The data obtained indicate that, during cyclical heat treatment of tin, structural changes acting in opposite ways on the stability of the gray tin centers evidently develop. The overall effect of the action depends on the total number and character of the heat-treatment cycles and the original properties of the surface. For a small number of cycles of heat treatment (2 to 6), the changes which arise increase preferentially the stability of the centers. Increasing the low-temperature holding time leads to the same effect, but for a large number of heat-treatment cycles structural changes with the opposite effect become stronger. There is an analogy between these processes and the activation and deactivation of impurities on crystallization of liquids.

During repeated recrystallizations of copper—aluminum and copper—tin alloys in [4], it was noted that martensite crystals formed repeatedly from the same sites, even though in intermediate cycles other crystals developed. It was also noted that superheating frequently led to a change in the observed microstructure. These experiments created the impression that there were a number of "prepared sites" in the sample, ready for developing centers of the new phase, and that, depending on the heat treatment, these might be activated or deactivated to various extents.

The examples considered above relate to polymorphic transformations taking place on cooling. Naturally, for transformations taking place on heating, we may expect deactivation of the sites ready for centers of the high-temperature phase, on holding at temperatures below the transformation point. Such data are to be found in [32], where the kinetics of the polymorphic transformation $NH_4NO_3(IV) \rightarrow NH_4NO_3(III)$ taking place on heating for $T > 32.3$°C are considered. It is shown, in particular, with maintenance of the temperature at 14°, supercooling substantially lowers the rate of the subsequent IV → III transformation at $T = 34.55$°C, as compared with first supercooling at 0.8°C. This conclusion was also confirmed by the authors in experiments where the transformation took place at 33.6°C. In other experiments it was shown that increasing the holding time at room temperature led to an increase in the time-lag before reaching the maximum transformation rate at 34.6°C, the actual value of this rate remaining practically unchanged.

From the point of view of the concepts of [25], if these are extrapolated to the case of phase transformations in solids on heating, the aggregate of the experiments in [32] indicates the conservation of phase III centers at small supercoolings (0.8 to 1.8°) and their disruption on supercooling to room temperature, the more severely, the greater the holding time at room temperature. This interpretation is also supported by the analysis of kinetic constants made in the same paper.

Data Obtained on Phase Transformations in Powders

A number of essentially new results touching both the role of the "prepared sites" in phase transformations in solids and the effect of thermal history on their kinetic aspects were obtained in experiments with powders.

Firstly, these experiments convincingly demonstrated the heterogeneous formation of new-phase centers. Thus it was shown in [33] during a study of polymorphic transformations in Ag_2SO_4 powders, that for a given temperature the rate of transformation decreased sharply with time. This fact, as well as the fact that at the given temperature there was a large amount of untransformed powder particles, the authors explained as being due to the heterogeneous generation of centers of the low-temperature phase.

In experiments on the martensite transformation in powders of iron—nickel alloys [34, 35], it was shown that, with diminishing grain size, the total amount of martensite phase in the transformed powders fell. It was concluded in [35] by way of explanation that, as the degree of dispersion of the powder increased, individual grains might no longer contain sites "prepared" for the generation of martensite crystals. It follows from the data of [34] that, as the particle size decreases, so also does the proportion of particles undergoing transformation, since the probability of finding "active" structural features in a given grain diminishes.

It is also possible that, in the case of the martensite transformation in powders, the effect of stresses accelerating the transformation [36, 40] is less evident. It is pointed out in [37] that the absence of tensile stresses accelerates the tempering of martensite in powder, as compared with a continuous sample.

Secondly, in experiments with powders of different dispersion, the fact of the distribution of "prepared sites" with respect to activity arises more obviously. Thus Johannson [33] indicates that, with falling grain size, the number of potential seeds ("prepared sites" in our terminology) becomes so small in the majority of crystals that the maximum transformation rate shifts in the direction of lower temperatures, i.e., less and less active parts begin to play the part of "prepared sites," and correspondingly greater supercooling are required for transforming the powders. An interesting fact in this connection is mentioned in [33], namely, the effect of the surrounding medium on the transformation in silver sulfate powder grains. To prevent mutual contact of these grains, Al_2O_3, TiO_2, and other powders were used as a neutral filler (80 vol. %). For Ag_2SO_4 grains of diameter 23 to 26 and 7 to 11 μ, the temperature at which the transformation rate was maximum did not depend on the filler, being respectively 288 and 277°C (the transformation point was 428°C). For smaller grains the temperature began to depend on the filler. Thus, for a grain size of 1.3 to 2.4 μ, the temperature was as follows for Al_2O_3 filler, 218; for TiO_2, 246; for ZrO_2, 263; and for graphite, 232°C. Thus the filler ceases to be neutral for contact with powder grains which are so small that most of them do not contain any "prepared sites." Naturally, this effect appears at large supercoolings. For still larger supercoolings (T = 160°C), these very fine powders give an especially sharp maximum of the transformation rate for all fillers; as noted in [33], this may be due to the homogeneous generation of centers of the low-temperature form. Since, as the supercooling increases, transformation occurs for crystallites containing less and less active "prepared sites", the proportion of grains undergoing transformation also increases with the supercooling. This increase was also observed in [34, 38], where, for completely different cases, similar S-shaped curves relating the proportion x of transformed grains to the temperature were obtained. Accordingly, the relationship between dx/dT and temperature has the form of a Gaussian curve and corresponds to the activity distribution of the "prepared sites" in transforming crystals of the same size. It was also established from observations on the $\alpha \rightleftharpoons \beta$ polymorphic transformations in powder samples of cristobalite [38] that the temperatures of the forward and reverse transformations in individual grains, distributed in the ranges 225 to 241 and 253 to 270°C, were mutually independent. This suggests that in the case in question the "prepared sites" for the generation of α and β phases were different.

Thirdly, new data on the effects of the thermal history of the samples were obtained from powders. Thus in [22] it was observed that, as the number of cycles of transformations $\alpha \rightarrow \beta \rightarrow \alpha$ in powdered tin increased, the rate of transformation increased or decreased according to the temperature at which the gray tin transformed into white. If this transformation took place at 25.6°C, then the subsequent transformation of white tin into gray was increasingly accelerated with increasing number of transformation cycles completed. If, however, the α-Sn $\rightarrow$ β-Sn transformations took place at 45 or 50°C, then the succeeding transformations were retarded. From our point of view, these experiments revealed the phenomena of activation and deactivation of the regions in the tin powder grains at which second-phase centers were generated. Thus on making the tin powder finer the effect of an increase in the transformation rate after 37 $\beta \rightarrow \alpha \rightarrow \beta$ transformation cycles was considerably less than after a smaller number of cycles (for a larger number of cycles the powder grains break up and may not contain regions active for the transformation). Such a weakening of the influence of heat treatment in the case of a finer powder was also noted for the polymorphic transformation of Ag_2SO_4 [33]. Here in particular it was shown that, if the powder sample were heated to 435°(superheating $\Delta T = 7°$), then on subsequent cooling the maximum transformation rate occurred at 376°C (activation), i.e., at a temperature 90° higher than for stronger heating (e.g., to 475°C: deactivation).* Moreover, on heating to relatively high temperature (> 500°C) the total area under the transformation curve falls,and this is apparently also connected with the deactivation of the "prepared sites." The effects mentioned are more strongly expressed in the case of coarser-grained Ag_2SO_4.

The fact mentioned in [34], that repeated casting of iron—nickel alloy pellets led to a fall in the number of pellets in which the martensite transformation took place, may also be explained by the deactivation of "prepared sites" at high temperatures. Finally, the effect of the temperature at which crystallization of cristobalite grains from the gel took place on the temperature characteristics of the $\beta \rightleftharpoons \alpha$ transformation of these grains was noted in [38].

The possibility of using the concept of the activation and deactivation of "prepared sites" for the phenomena of the stabilization and destabilization of austenite during martensite transformations is discussed in [31]. As shown in [31], a study of these phenomena in powder samples could provide useful information as to their nature.

It should be noted that, although it would appear that the phenomena of the activation and deactivation of "prepared sites"during transformations in solids are widespread, the physical essence of the processes taking place during activation and deactivation is not clear. In studying the crystallization of liquids [39], the activation of impurities was explained by the formation of "molecular contact" on the ingrowth of an impurity crystal into the melt. Deactivation was explained by the flattening of pores and cracks in the insoluble particles, in which the generation of centers took place on submitting the system to cyclic heat treatment including recrystallizations. The treatment of such phenomena on cyclic recrystallization in liquids (given by Danilov) already seems rather complex and difficult to submit to experimental proof. Naturally, similar phenomena taking place on transformations in the solid state are far more complex, since the solid state involves the considerable part played by stresses, which are not present in the crystallization of liquids.

Conclusions

1. In the overwheleming majority of cases of phase transformations in solids, the generation of centers for the new phase takes place at certain "prepared sites." These may be various structural defects: included particles of other phases (insoluble impurities), grain boundaries, dislocations, and other nonequilibrium deviations from structural regularity.

2. There is an activity distribution of the "prepared sites," and this leads to an increase in the number of centers of the new phase on increasing the supercooling (or superheating).

---

* The author explains this effect of slight superheatings by saying that the low-temperature phase seeds are thermodynamically stable even above the transformation temperature. This is yet another confirmation of the possibility of extrapolating the conclusions of [25] to the case of phase transformations in solids.

3. The effect of the thermal history of samples on the kinetics of phase transformations in solids, known from experimental data, may be explained by extending Kazachkovskii's concept of the possibility of preserving second-phase crystallites in micropores of insoluble impurities, within a certain temperature range of the transformation point, to the case of phase transformations in solids, either on heating or cooling.

4. A number of important new results concerning both the role of the "prepared sites" in phase transformations in solids and the effect of the thermal history of samples on the kinetics of these transformations, were obtained in experiments with powders. In these experiments the heterogeneous formation of new-phase centers was convincingly demonstrated. With the use of powders of different dispersions, the activity distribution of the "prepared sites" and the effect of heat treatment on this activity are indeed more clearly discernible.

Literature Cited

1. D. Turnbull, Collection: Impurities and Defects [Russian translation from the English], Editor, B. N. Finkel'shtein (Metallurgizdat, Moscow, 1960), p. 141.
2. E. Scheil and A. Lange-Weise, Arch. Eisenhüttenwesen, 11:93 (1937).
3. Yu. V. Grdina and L. A. Bondar' Izv. Vuzov, Chern. Metal., 4:73 (1959).
4. I. A. Arbuzova and L. G. Khandros, Questions of the Physics of Metals and Physical Metallurgy, No. 14 (1962).
5. M. M. Chertok, Zh. Tekhn. Fiz., 5(4):711 (1935).
6. B. R. Butcher and A. N. Holden, Progr. Nucl. Energy, V. 2:419 (1959).
7. E. C. Bain, Trans. Amer. Soc. Steel Treat., 20:385 (1932).
8. N. A. Goryunova, Gray Tin (Master's Dissertation), Leningrad State University (1951).
9. A. I. Bykhovskii, Fiz. Metal. i Metalloved., 6(3):487 (1958).
10. A. I. Bykhovskii, Dokl. Akad. Nauk SSSR, 139(3):637 (1961).
11. R. R. Rogers and J. F. Fydell, J. Electrochem. Soc., 100(4):161 (1953).
12. N. A. Goryunova, Dokl. Akad. Nauk SSSR, 75(1):51 (1950).
13. G. Donze and R. Faivre, Compt. Rend. Acad. Sci., 245(25):2277 (1957).
14. G. Donze and R. Faivre, Compt. Rend. Acad. Sci., 246(26):3619 (1958).
15. W. Fraenkel and W. Goez, Z. Anorg. Allgem. Chem., 144(1−2):45 (1925).
16. R. D. Nelson, Trans. Am. Soc. Metals, 51:677 (1959).
17. Yu. F. Frichshe, Memoires de l' Acad. Imper. Sci. St-P, VII, Ser. 15, No. 5, (1870).
18. E. Cohen and C. V. Eijk, Z. Phys. Chem., A., 30:601 (1899).
19. E. Cohen, Z. Phys. Chem., A., 35:588 (1900).
20. G. Tammann and K. L. Dreyer, Z Anorg. Allgem. Chem., 199:97 (1931).
21. W. G. Burgers and L. J. Green Disc. Faraday Soc., No. 23, 183, (1957); general discussion, p. 220
22. E. A. Cohen and K. W. A. von Lieshaut, Z. Phys. Chem., A., 173:1 (1935).
23. V. I. Danilov and V. E. Niemark, Zh. Eksper. i Teor. Fiz., 7(9−10):1168 (1937).
24. V. I. Danilov, Problems of Physical Metallurgy and the Physics of Metals [in Russian]. Metallurgizdat, Moscow, No. 1, 7, 1949.
25. O. D. Kazachkovskii, Collection of Scientific Papers of the Metallophysics Laboratory, Akad. Nauk Ukr. SSR, No. 76 (1948).
26. D. Turnbull, Acta Met., 5(9):502 (1956).
27. G. Freeman and G. J. Dienes, J. Appl. Phys., 26(6):652 (1955).
28. E. Cohen, Z. Phys. Chem., A., 115:151 (1925).
29. J. N. Brönsted, Z. Phys. Chem., A., 131(5−6):366 (1928).
30. A. F. Benton and R. D. Cool, J. Phys. Chem., 35(6):1762 (1931).
31. A. I. Bykhovskii, Physical Metallurgy and Heat Treatment (Mashgiz, 1961), p. 317.
32. B. V. Erofeev and N. N. Mitskevich, Zh. Fiz. Khim., 24(10): 1235 (1950).
33. G. Johnannson, Arkiv for Kemi, 8, No 4, 33, 1955.
34. R. E. Cech and D. J. Turnbull, J. Metals 8(2):124 (1956).

35. G. V. Kurdyumov and L. G. Khandros, Questions of the Physics of Metals and Physical Metallurgy, Izv. Akad. Nauk Ukr. SSR, (9):3 (1959).

36. A. P. Gulyaev and V. D. Zelenova, Fiz. Metal. i Metalloved., 6 (5): 945 (1958).

37. A. P. Gulyaev and V. D. Zelenova, Fiz. Metal, i Metalloved., 6 (5): 936 (1958).

38. O. Krisement and G. Trömel, Z. Naturforsch., 14a, No. 7:685; No. 10:912

39. V. I. Danilov, Collection of Scientific Papers of the Metallophysics Laboratory (Izd. Akad. Nauk Ukr. SSR, 1948), p. 95.

40. E. S. Machlin and M. Cohen, J. Metals, No. 9:755 (1951).

SOME QUESTIONS ON THE KINETICS OF PHASE TRANSFORMATIONS

L. N. Aleksandrov

Phase transformations, including various forms of precipitation from supersaturated solid and liquid solutions, crystallization, polymorphic transformations, and recrystallization processes, take place by the ordinary mechanism of the generation of centers and their subsequent growth. The relation between the proportion of transforming volume η and the transformation time t determines the kinetics of the process. A kinetic equation allowing for the statistical character of the overlapping of growing centers of the new phase was obtained by Kilmogorov [1]. This equation, valid on the assumption that grains grow uniformly in all directions or for the case of identically oriented crystals or arbitrary form, and improved by Rodigin [2] so as to obtain a finite t for $\eta = 1$, does not reveal the physical characteristics of the kinetics of specific processes, since it gives a general statistical-probability solution for the problem of the transformation time. More complete consideration of the kinetics of phase transformations on the basis of the equations of Avrami [3], Mehl and Johnson [4], and Kazeev [5] requires the analysis of time-dependence to include not only the rates of generation and growth but also a number of auxiliary parameters such as relaxation time, so that the majority of theoretical and experimental investigations of the kinetic aspect [6-12] are based on Kolmogorov's equation:

$$\eta(t) = \frac{V_0 - V(t)}{V_0} = 1 - \exp\left[-\int_0^t I(\xi) v_0(t-\xi) d\xi\right], \tag{1}$$

where V_0 is the original volume, $V(t)$ the untransformed volume, $I(\xi)$ the rate of generation of the new phase, $v_0(t-\xi)$ the volume of the growing new-phase center. For slight supercoolings of the system, dilute solutions, and small degrees of transformation, the growth of crystals proceeds freely, and simpler conditions which do not allow for the overlapping of the growing centers of the new phase may be applied. A kinetic equation of this form was first obtained by Sachs for the case of the crystallization of a supercooled liquid, and his solution for a three-dimensional transformation appears in [13]. In general, this kinetic equation is an integral equation of the second kind:

$$V_0 - V(t) = \int_0^t I(\xi) V(\xi) v_0(t-\xi) d\xi \tag{2}$$

and it follows from this that the number of centers being formed at the instant ξ will be $I(\xi) V(\xi)$.

Application of operational calculus enables us to solve equation (2) for various types of transformation kinetics.

Let us consider kinetic equation (2) for the steady-state generation and growth of centers at a rate which is constant for a given temperature. Here $I(\xi) = \text{const} = I_0$; $v_0(t-\xi)$ for the case of linear (needle), two-dimensional (disk), and three-dimensional (spherical) precipitates equals respectively

$$v_0(t-\xi) = \alpha_i c^i (t-\xi)^i, \quad i = 1, 2, 3. \tag{3}$$

Here α_i is the form factor which in the simplest cases considered is respectively $\alpha_1 = 2S$; $\alpha_2 = \pi d$; $\alpha_3 = 4/3\pi$, where d is the disk thickness and S the cross section of the needle. For other types of precipitate the values of α_i vary.

Let us proceed to the relative volume in (2), and after substituting (3) we carry out an integral transformation [14]. For the image we obtain

$$\overline{V(p)} = \frac{p^i}{i\,\alpha^i + p^i}, \quad i = 2,\ 3,\ 4. \tag{4}$$

Transforming to the original, we obtain the solution of equation (2). For a linear growth of the centers

$$V(t) = V_0 \cos (2I_0 cS)^{1/2}\, t. \tag{5}$$

For two-dimensional precipitation,

$$V(t) = \frac{1}{3}V_0 \left\{ \exp\left[-(\pi I_0 c^2 d)^{1/3}\, t\right] + 2\exp\left[(\pi I_0 c^2 d)^{1/3}\, t\right] \times \right. \tag{6}$$
$$\left. \times \cos \frac{\sqrt{3}}{2}(\pi I_0 c^2 d)^{1/3}\, t \right\}.$$

For three-dimensional,

$$V(t) = V_0 \cos (2\pi I_0 c^3)^{1/4}\, t\, \mathrm{Ch}\,(2\pi I_0 c^3)^{1/4}\, t. \tag{7}$$

In the case of the diffusion mechanism of growth (time-dependence of the form $t^{-i/2}$), the solution of the kinetic equation is obtained by applying to the expression $f(t)$ in (5), (6), (7) the "root-taking" transformation, which gives

$$\frac{1}{\sqrt{\pi t}} \int\limits_0^\infty e^{-z^2/4t}\, f(z)\, dz.$$

The region in which these solutions may be applied to the study of the kinetics of three-dimensional transformations is indicated in [13]. For the case of two-dimensional generation and growth this region is shown in Fig. 1: Curve 1 is calculated from equation (6) for free growth, and curve 2 from equation (8), derived from (1) and allowing for the superposition of centers:

$$V(t) = V_0 \exp\left(-\frac{\pi}{3}I_0 c^2 t^3\right). \tag{8}$$

For a degree of transformation of 0.2 to 0.3, the kinetics are equally well described by either equation, so that the simple relations of the form (5) or (7) may be used.

Another simple case of kinetics is crystallization or precipitation of prepared centers, the number N of which neither rises nor falls in the transformation process; this takes place in the dispersion hardening of alloys,

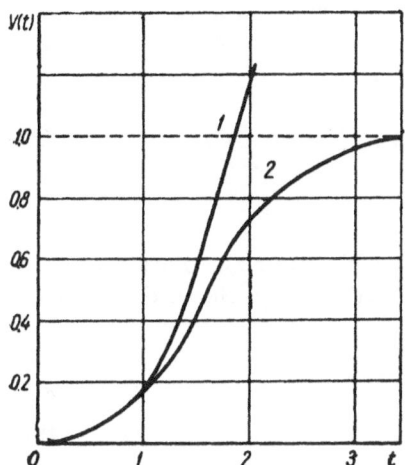

Fig. 1. Curves relating the proportion of transformed volume to the transformation time for two-dimensional generation and growth, assuming 1) free growth of centers, and 2) superposition.

selective recrystallization, and the decomposition of supersaturated solid solutions in semiconducting systems with a small concentration of the dissolved component [15] in the presence of impurities and dislocations. Not allowing for the overlapping of centers, we obtain for V(t)

$$V(t) = V_0 - N v_0(t). \tag{9}$$

The kinetic equation (9) may be obtained from (2) if we assume that there is a connection between N, I_0, and t of the form $N = I_0 t$. This case is considered in [1] as singular and solved by a limiting transformation which fails to reveal the physical essence of the process. Formal application of this scheme to the two-stage transformation in the papers of Meleshko [16] and Zaremba [17], in which the generation rate was neglected at one stage and only the prepared centers counted, leads to errors in estimating the kinetic parameters.

Let us consider the kinetics of a transformation taking place successively at the temperatures T_1 and T_2. If the generation rates are respectively I_1 and I_2, the growth rates c_1 and c_2, and the transformation time at the first temperature τ, then in the general case we obtain from equation (1)

$$\frac{V(t)}{V_0} = \exp\left[-\int_0^\tau I_1(\xi) v_0(t-\xi) d\xi - \int_\tau^t I_2(\xi) v_0(t-\xi) d\xi\right]. \tag{10}$$

For steady-state two-dimensional generation, upon integrating we obtain

$$\frac{V(t)}{V_0} = \exp\left[-\frac{\alpha_2}{3} I_1 c_1^2 \tau (t^2 + t\tau + \tau^2) - \frac{\alpha_2}{3} I_2 c_2^2 (t-\tau)^3\right]. \tag{11}$$

In the case of three-dimensional precipitation,

$$\frac{V(t)}{V_0} = \exp\left\{-\frac{\alpha_3}{4} I_1 c_1^3 \tau \left[t^2 + (t-\tau)^2\right](2t-\tau) - \right.$$

$$\left. -\frac{\alpha_3}{4} I_2 c_2^3 (t-\tau)^4\right\}. \tag{12}$$

Since generation takes place continuously, in equations (11) and (12) clearly $N = I_1\tau$. Analysis of the kinetic equations obtained after substituting N shows that these correspond to those taken in [16, 17] for $\tau = 0$, but since in these investigations τ exceeds 0.5t, the approximation taken must be regarded as very coarse.

In conclusion, we must note that the kinetics of transformations may be considered from general physical concepts by analyzing the size-distribution function of the centers z(x, t) [18, 19]. This function is determined by solving the kinetic equation

$$\frac{\partial z}{\partial t} = \frac{\partial}{\partial x}\left(D_1 \frac{\partial z}{\partial x} - D_2 z\right), \tag{13}$$

in which D_1 and D_2 are kinetic coefficients corresponding to the fluctuation variation in the sizes of the centers as a result of "diffusion" along the axis of dimensions and directional growth ($D_2 = v$). For free growth of centers, the transformed volume is determined by direct integration from the size corresponding to critical:

$$V(t) = V_0 - \int_{x_{\text{кр}}}^{\infty} z(x, \ t) \, v_0(x) \, dx. \tag{14}$$

Allowing for the overlapping of the centers considerably complicates the form of functions $z(x, t)$ and $v_0(x)$; carrying out the corresponding calculations, however, leads to equations analogous to those obtained from (1).

As regards the kinetic equation

$$V(t) = V_0 \exp \left[- \left(\frac{t}{\tau_0} \right)^n \right], \tag{15}$$

used in [5], [10], [20], and others, as remarked earlier, this is equally valid with equation (1), although it fails to reflect the elementary mechanisms of the process, as it only describes the kinetics by formal statistics.

The development of the theory of the kinetics of phase transformations must be directed both toward a more profound elucidation of the physical content of the special cases in the Kolmogorov theory and toward discovery of the form of the time-dependence of the size-distribution function of new-phase centers.

Literature Cited

1. A. N. Kolmogorov, Statistical Theory of the Crystallization of Metals, Izd. Akad. Nauk SSSR, Ser. Mat., (3):355 (1937).
2. V. N. Rodigin, Theory of Crystallization, Zh. Tekhn. Fiz., 22 (8):1356 (1952).
3. M. Avrami, Kinetics of Phase Change, J. Chem. Physics, 7, No. 12, 1103 (1939); 8, No. 2, 212; (1940); 9, No. 2, 177, (1941).
4. W. A. Johnson and R. F. Mehl, Trans. AIME, 135: 416 (1939).
5. S. A. Kazeev, Kinetics in Application to Physical Metallurgy (Oborongiz, 1956).
6. N. N. Sirota, Analytical Expression of the Kinetic Curves of Phase Transformation, Dokl. Akad. Nauk SSSR, 36 (6):192 (1942).
7. Yu. V. Grdina and L. A. Eliseeva, Crystallization Equations, Dokl. Akad. Nauk. SSSR, 109 (3):566 (1956).
8. L. N. Aleksandrov and B. Ya. Lyubov, Effect of Alloying on the Kinetics of the Pearlite Transformation, Fiz. Metal. i Metalloved., 8 (2):216 (1959).
9. L. N. Aleksandrov and B. Ya. Lyubov, Theoretical Analysis of the Kinetics of Decomposition of Super-saturated Solid Solutions, Usp. Fiz. Nauk, 75 (1):116 (1961).
10. J. S. Kirkaldy, Theory of diffusional growth in solid—solid transformation, in: Decomposition of Austenite by Diffusional Processes (New York, 1962).
11. I. W. Cahn, The kinetics of grain boundary nucleated reactions, Acta Metallurgica, 4 (5):449 (1956).
12. J. E. Burke and A. D. Turnbull, Recrystallization and Grain Growth, Progr. Metal Physics, 3 (1952), p. 220; Uspekhi fiziki metallov, Vol. 1, Metallurgizdat, 1956).
13. G. S. Zhdanov, Solid State Physics (Izd. MGU, 1962).
14. A. I. Lur'e, Operational Calculus and its Application to Problems of Mechanics (Gostekhizdat, 1950).
15. B. I. Boltaks, Diffusion in Semiconductors (Fizmatgiz, 1961).
16. L. O. Meleshko, Crystallization of Supercooled Liquids in a Field of Ultrasonic Waves, Inzh. Fiz. Zh., (4):4 123 (1961). Summaries of contributions to the All-Union Conference on the Thermodynamics and Kinetics of Phase Transformations (Minsk, 1962).

17. V. G. Zaremba, Kinetics of Heterogeneous Crystallization on Formation of an Organic Bar, Inzh.-Fiz. Zh. Akad. Nauk Belorussk. SSr, 4(5):74 (1961).

18. Ya. I. Frenkel', Introduction to the Theory of Metals Fizmatgiz (1959).

19. B. Ya. Lyubov, Nonsteady-State Generation Rate of New-Phase Centers at Large Supercooling Velocities, Dokl. Akad. Nauk SSSR, 91 (2):245 (1953).

20. O. Krisement and F. Wever, The Bainite Reaction in High-Carbon Steel: The mechanism of phase transformations in metals, London, 1956, p. 253; Phase Transformations in Steel [Russian translation], Metallurgizdat, 1961.

EFFECT OF THE DIFFUSION OF VACANCIES ON THE GROWTH KINETICS OF CRYSTALS DURING THE RECRYSTALLIZATION AND DECOMPOSITION OF SUPERSATURATED SOLID SOLUTIONS

L. N. Aleksandrov

The kinetics of transformations in the solid state are affected by the degree of imperfection of the crystal structure, which determines the mobility of the atoms. Structural defects such as vacancies and dislocations are either already present in the initial state of the solid, if this has been subjected to working, irradiation, or quenching from high temperatures, or else develop as new-phase centers are formed on heating or cooling. Studies of the thermodynamics and kinetics of recovery on annealing metals and alloys [1, 2] have shown that up to the temperature of the onset of recrystallization the change in the macroscopic properties and evolution of energy are caused by processes associated with highly mobile point defects, diffusion and pairing of vacancies, formation of vacancy accumulations, migration of vacancies, and dislocations. The concentration of vacancies tends to the equilibrium value

$$C_\mathrm{p} = \exp\left(-\frac{Q}{RT}\right),$$

(1)

where Q is the energy of formation of a mole of vacancies.

The interaction of vacancies with dislocations and impurity atoms, however, ensures the maintenance of an abundant concentration of vacancies up to the high temperatures at which the displacement and climbing of dislocations begins, leading first to polygonization and then to recrystallization. Each dislocation in the crystal is surrounded by a field of elastic stresses, so that the equilibrium concentration of vacancies interacting with dislocations is different in different parts of the solid. The interaction of vacancies and dislocations determines the diffusion mobility of the atoms, and hence the mechanical properties of the crystal. Thus creep is associated with the climbing of dislocations, which is caused by the diffusion of vacancies to the axis of the dislocations under the influence of a vacancy concentration gradient produced by the gradient of elastic stresses from the dislocation [3]. The diffusion of vacancies takes place between dislocations of different signs, and the diffusion of dissolved atoms in solid solution takes place among vacancies and interstitial sites.

The accelerating action of stresses in the crystal on diffusion may be explained by the formation of a large number of excess vacancies on restoring the regularity of the lattice. The inhomogeneous distribution of vacancies in a deformed crystal is caused by the inhomogeneous distribution of stresses, since the vacancy concentration depends linearly on the degree of lattice distortion [4]. The retention of excess vacancies on heating is caused by the presence of disoriented gains, phase-separation boundaries, impurity atoms, and pinned dislocations.

Systems of moving dislocations generate excess vacancies in the region of the recrystallization temperature [5], and the concentration of these vacancies is naturally higher in places of increased dislocation density, i.e., in the more stressed parts.

The inhomogeneous distribution of vacancies leads to the appearance of diffusion flows which level out the concentration and superimpose themselves on the motion of the dislocations.

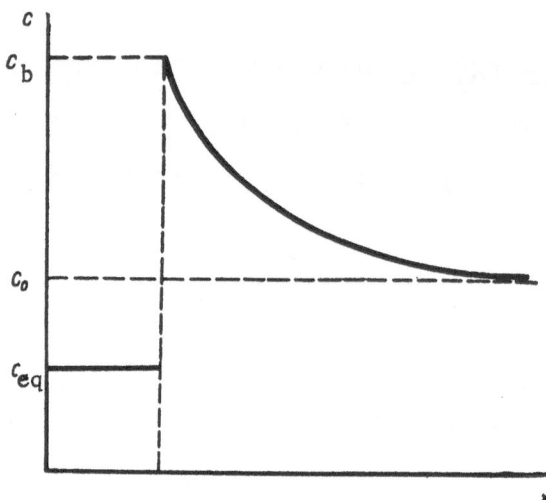

Fig. 1. Vacancy distribution in a crystal during recrystallization.

Hence we should take account of the role played by vacancies in transformations at temperatures higher than those usually associated with vacancy migration.

In this paper we consider the effect of the mechanism of vacancy diffusion on the kinetics of grain growth during the recrystallization and decomposition of supersaturated solid solutions.

Recrystallization in deformed solids on heating takes place as a result of the inhomogeneous distribution of stresses, which makes it possible to reduce the free energy of the system by growth of the unstressed regions. As recrystallization proceeds, distortions are removed; the concentration of vacancies falls in the recrystallized parts and rises sharply near crystal boundaries, which act as vacancy sinks.

A difference in vacancy concentration is created between the original deformed part, the recrystallized part, and the boundary, and the vacancy diffusion flows which develop accompany the growth of the recrystallizing grain. The vacancy distribution in the crystal during recrystallization is depicted schematically in Fig. 1. To calculate the vacancy concentration C_0 in the distorted part and C_b at the crystal boundary, we may use the relation of [6], which allows for the additional distortion energy E from the vacancies. The vacancy concentration at the boundary is determined by the relation

$$C_b = \frac{C_{eq}\exp\left(\dfrac{Q-E}{RT}\right)}{1-C_{eq}+C_{eq}\exp\left(\dfrac{Q-E}{RT}\right)}. \qquad (2)$$

The quantity E is estimated from the free energy of the grain surface σ and the atomic radius r_a. If N is Avogadro's number, then

$$E = \frac{1}{C_b} N \pi r_a^2 \sigma. \qquad (3)$$

The concentration in the original deformed part C_0 will be determined by the total distortion energy for one mole of vacancies. In this case

$$E = \frac{v_0 \Delta F_0}{C_0}, \qquad (4)$$

where v_0 is the specific volume of the metal, and ΔF_0 is the change in free energy with the formation of unit recrystallized volume.

In a recrystallizing metal, vacancies may be considered as impurity atoms of a second component. Hence the rate of motion of a recrystallized grain boundary, assuming that the boundary is formed by the outflow of excess vacancies from the grain boundary with the equilibrium concentration C_{eq} into the deformed region with concentration C_0, may be calculated by solving the vacancy diffusion equation by the method used earlier in calculating the rate of growth of precipitates in two-component systems [7]. The boundary conditions of the equation are preserved. In the case of a spherical grain, the mean size at instant t is determined by the vacancy diffusion coefficient D_0:

$$2\rho = 4\beta \sqrt{D_0 t},\tag{5}$$

and the velocity is

$$v = \frac{d\rho}{dt} = \frac{2\beta^2 D_0}{\rho} = \frac{\beta \sqrt{D_0}}{\sqrt{t}}.\tag{6}$$

The value of parameter β is determined by the original vacancy distribution in the deformed metal.

For growth of the crystal in the form of a plate, its edge may be considered as having the form of a parabolic cylinder. Then the radius of curvature ρ_0 at the vertex of the parabola and the rate of motion of the edge of the plate are connected with parameter p and the vacancy diffusion coefficient by the relation

$$v = \frac{p D_0}{\rho_0}.\tag{7}$$

For rapid establishment of the steady state at the front of the lamellar crystal, parameters p and ρ_0 remain constant, and growth must take place at a rate which is constant for given temperature and degree of deformation.

Experimental investigations of the kinetics of recrystallization have shown [8−10] that even in single-component systems the process takes place over a certain range as a diffusion transformation in which the relation between mean grain size and time is of the form (6), or

$$2\rho = K \ln t + D_0.\tag{8}$$

Whereas for a spherical grain the time-variation of the boundary displacement rate is characteristic for diffusion-limited processes, in the case of a hyperbolic fall in velocity for lamellar precipitation

$$\frac{d\rho}{dt} = \frac{K}{t}\tag{9}$$

and additional considerations regarding the nature of the time-variation of the radius of curvature of the boundary are necessary. Let us suppose that ρ_0 rises on a linear law

$$\rho_0 = bt.\tag{10}$$

This kind of change in boundary shape occurs in the growth of large-angle grains; the grain becomes rounded and further growth takes place in accordance with (6).

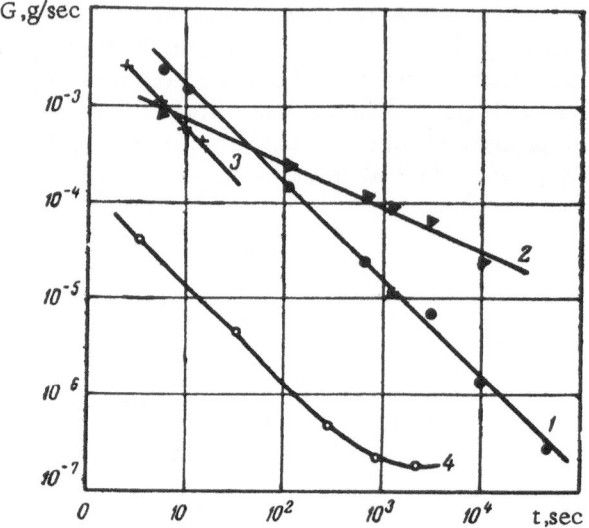

Fig. 2. Crystal growth rate in molybdenum on recrystallization; computed values, curves 1 and 2; experimental, curves 3 and 4.

As an example of calculating grain-growth rates during recrystallization by the vacancy diffusion mechanism, we considered molybdenum crystals. The value of b was estimated from experimental data ($\rho_0 = 2 \cdot 10^{-3}$ cm at t = 100 sec).

The vacancy diffusion coefficient in the bcc lattice of molybdenum is taken from the results of [11]:

$$D_0 = 10^{-2} \exp(-40000/RT) \text{ cm}^2/\text{sec}$$

Parameter β in (6) is calculated from the equation [7]

$$\frac{C_b - C_0}{C_b - C_{eq}} = 2\beta^2 (1 - \beta \sqrt{\pi} e^{\beta^2} \text{erfc } \beta) = F(\beta). \tag{11}$$

For the growth of a plate, the edge conditions for parameter p in (7) give the equation

$$\frac{C_b - C_0}{C_b - C_{eq}} = \sqrt{\frac{\pi p}{2}} e^{p/2} \text{erfc} \sqrt{\frac{p}{2}} = \varphi(p). \tag{12}$$

For C_{eq} at T = 1500°C we obtain from (1) $3 \cdot 10^{-5}$, since Q = 40 kcal/g·atom. Let us find C_b and C_0 from (2) —(4). Taking σ = 5 ergs/cm², $\Delta F_0 = 1$ cal/cm³ [12], $v_0 = 9.45$ cm³/g·atom, $r_a = 1.36$ A, we have $C_b = 2 \cdot 10^{-3}$ and $C_0 = 4 \cdot 10^{-4}$.

This gives for F(β) and φ(p), 0.82. The corresponding values of β and p are 2.5 and 3.0. The graph of function φ(p) for large p is given in [13]. Function F(β) for β≫1 reduces to the form

$$F(\beta) = 1 - \frac{1,3}{2\beta^2} . \tag{13}$$

The results of calculating the molybdenum crystal growth rate during recrystallization (assuming the vacancy diffusion mechanism to be the decisive factor) from (7) and (10) for plates and from (6) for spherical grains are shown in Fig. 2. The experimental data represented by curves 3[9] and 4[10] are lower than calculated.

Analysis of the results leads to the conclusion that the redistribution of vacancies on recrystallization can limit the boundary-displacement rate of the growing crystal in the early stages of growth.

As the front of the growing crystal straightens there is more satisfactory agreement between the experimental and calculated velocities.

Allowing for the effect of boundary curvature on the vacancy concentration C_b does not change the nature of the conclusions. In fact, there is a surplus concentration of vacancies (impurity atoms) on a convex surface as compared with that on a plane boundary, owing to the additional work of formation of this surface:

$$C_b^\rho = C_b^\infty \exp\left(\frac{2\sigma v_0}{\rho\,RT}\right). \tag{14}$$

For $\rho \leq (2\sigma/RT)v_0$ the number of vacancies may increase 10 to 100 times, and if the flow of vacancies were to pass to the grain boundary (for example, on precipitation of a center with a more distorted lattice than the original matrix), then on either raising or lowering ρ the growth rate would approach zero: $C_b^\rho = C_0$, and the vacancy-concentration gradient would vanish. In the case of recrystallization (Fig. 1), reduction in the curvature of the grain boundary leads to an increase in the concentration difference $C_b^\rho - C_0$ and hence to an acceleration in the redistribution of vacancies.

Figure 3 shows schematically the variation of the growth rate of a new-phase center with the radius of curvature of the boundary. The effect of boundary curvature at 10^{-6} cm may evidently be neglected in the majority of cases (the numerical data of Fig. 3 relate to the growth of a carbide lamella in a Fe—C solid solution at 600°C, considered in [13]; the part of vacancies is played by impurity atoms, but the character of the relationship is preserved).

The participation of vacancies in the decomposition of solid solutions was discussed in [5]. Using the precipitation of Cu from Al—Cu alloy as an example, the accelerating role of vacancies and their retention at high temperatures was demonstrated. Thus, as a result of deformation the excess concentration of vacancies at 200°C was retained and estimated as equilibrium at 500°C ($2 \cdot 10^{-5}$). The acceleration of precipitation from a supersaturated solid solution after quenching as a result of the formation of excess vacancies was shown in [15].

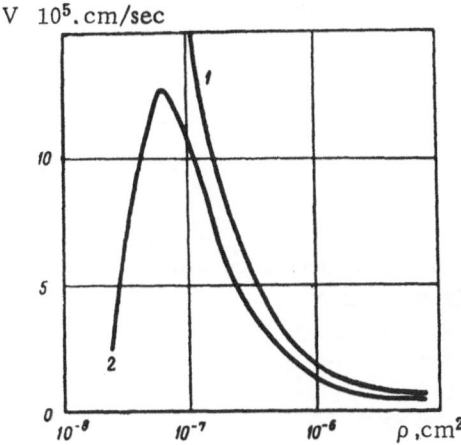

Fig. 3. Growth rate of a new-phase center as a function of the radius of curvature of the boundary (1) without and (2) with allowance for the effect of curvature on the equilibrium vacancy concentration at the boundary.

The mechanism of the effect of vacancies on the kinetics of precipitate growth during the decomposition of supersaturated solid solutions is different in substitution and interstitial solid solutions, and is determined by previous mechanical and heat treatment or irradiation. During decomposition, another distribution is superimposed on the original field of vacancies in solid solution; this arises during growth of the new-phase grains as a result of volume and concentration stresses. The developing stress field either retards the growth of the crystal, making it thermodynamically impossible (which takes place for diffusionless phase transformations), or accelerates growth, if the decisive mechanism is the diffusion redistribution of atoms. One of the factors causing an acceleration of growth is the development of a vacancy concentration in the stress field through which diffusion takes place. The overall flow of vacancies increases on account of those generated during the displacement and formation of dislocations, and this strengthens the diffusion counter-flow of atoms among the vacancies.

During the growth of a precipitate containing more than the original amount of the dissolved component in substitution solid solutions, vacancies diffusing into the depths of the solid mother solution promote the restoration of the original concentration at the boundary of the precipitate, and grain growth is accelerated. On precipitation of a new phase containing an amount of the dissolved component smaller than that in the original solid solution, an excess concentration not only of vacancies but also of atoms of the second component is created at the boundary. The migration of vacancies from the boundary of the new phase produces a flow of atoms of the dissolved component counter to the main flow from the boundary, and hence the growth of the center is retarded.

It should be noted that in the first of the cases considered, for a small radius of curvature of the boundary, the concentration gradient of dissolved atoms may vanish, as shown in Fig. 3, and the growth rate will diminish independently of the accelerating effect of the vacancy diffusion.

In interstitial solid solutions, diffusion of the dissolved component takes place through the interstitial sites, so that vacancy migration should have no effect on the growth rate of the new-phase center in either case. However, the excess concentration of vacancies at the boundary and their displacement from the boundary accelerate the self-diffusion of the main component, as a result of which the stresses at the boundary of the new-phase center diminish. Hence the accelerating action of the stresses diminishes, and so therefore does the growth rate of the precipitate.

An estimate of this diminution is made on the basis of the proportionality existing in a certain volume between the self-diffusion coefficients, vacancy concentrations, and stresses [16]:

$$\frac{D'}{D} = \frac{C'}{C} = \frac{\sigma'_r}{\sigma_r}. \tag{15}$$

The loss of vacancies is determined under the assumption of their equilibrium distribution at a spherical center of the new phase. The vacancy-concentration gradient at a grain boundary of radius ρ is

$$\left(\frac{\partial C}{\partial r}\right)_\rho = \frac{C_b - C_\infty}{\rho}, \tag{16}$$

where C_∞ is the vacancy concentration in the original solid solution. We may suppose that $C_\infty = C_{eq}$ if vacancies created by quenching or irradiation are not retained in the original lattice. In the opposite case $C_\infty = C_0$, as for recrystallization growth.

The flow of vacancies q through 1 cm^2/ sec is

$$q = D_0 \left(-\frac{\partial C}{\partial r}\right)_\rho. \tag{17}$$

Hence from (15) and (16) the decline in the growth rate of a spherical center, in accordance with (6), on discharge of excess vacancies from the boundary, will be determined by the relation

$$\frac{\beta_\rho}{\beta_\infty} = \alpha \frac{\sqrt{C_b - \frac{D}{\rho}(C_b - C_\infty)}}{\sqrt{C_\infty}}, \tag{18}$$

where we find the parameter α from the connection between the vacancy concentration and the value of β with and without consideration of the stresses acting.

Let us apply the considerations discussed to an Fe−C solid solution in which a ferrite grain is growing. At 720°C and an initial carbon concentration of 0.5 wt. %, by solving the diffusion equation with due allowance

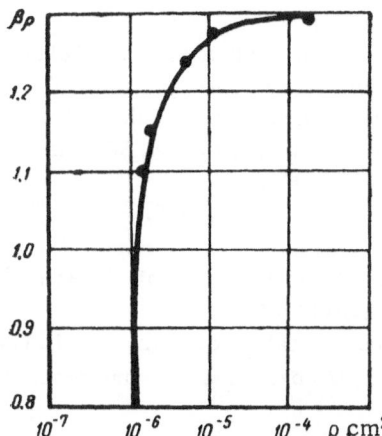

Fig. 4. Effect of vacancy diffusion from the grain
boundary on the growth rate (β_ρ) of a ferrite center
as a function of grain radius.

for the elastic stresses arising, we obtain $\beta_{str} = 1.32$. In the unstressed solid solution $\beta_\infty = 0.8$ [17]. From (1) and
(2) $C_b = 4 \cdot 10^{-3}$, $C_{eq} = 10^{-5}$. The activation energy of the formation and displacement of vacancies in iron
is 25 kcal/g·atom. Experimental data given in [18] indicate the stress relief taking place. Results of calculating the grain-growth rate (β_ρ), allowing for the diffusion of vacancies from the boundary, appear in Fig.
4; the retarding effect of the excess vacancies appears for centers of small size.

The relative stability of the positions of atoms in the lattice during the recrystallization and decomposition
of substitution solid solutions indicated in [19] does not contradict our main conclusions, since the vacancies
appearing are rapidly filled by atoms, the distance traveled by each vacancy is not large, and the regions in
which the effects of individual vacancies on the kinetics of the process appear may be of the order of atomic
dimensions.

Conclusions

1. Our analysis of the effect of vacancy diffusion on the kinetics of crystal growth during recrystallization
and decomposition of supersaturated solid solutions has shown that, despite the high mobility of vacancies
compared with dislocations and impurity atoms, the continuous variation in their distribution and concentration
during the transformation causes acceleration in the growth of precipitates in single-component deformed
systems and either acceleration or retardation of precipitation in solid solutions.

2. Since the impurity atom concentration usually approaches or exceeds the concentration of vacancies
created on plastic deformation, it is the impurity atoms rather than the vacancies which play a dominant role
in the kinetics of transformations in solid solutions.

Literature Cited

1. H. M. Clarebrough, M. E. Hargreaves, and G. M. West, Proc. Roy. Soc., 232A:252 (1955).
2. M. B. Biver, Collection: Creep and Recovery (Metallurgizdat, 1961).
3. B. Ya. Pines, Ukr. Fiz. Zh., 76 (3); 51 (9) (1952).
4. S. D. Gertsriken and M. M. Novikov, Visnik KDU 1:63 (1958).
5. F. Seitz, Advances in Physics, 1 (1):3 (1952).
6. V. T. Shmatov and A. V. Grin', Fiz. Metal. i Metalloved., 12 (4):600 (1961).
7. L. N. Aleksandrov and B. Ya. Lyubov, Usp. Fiz. Nauk, 75 (1):117 (1961).

8. P. A. Beck, Z. Metallkunde, 52 (1):13 (1961).

9. N. V. Grevtsev and G. N. Klebanov, Izv. Akad. Nauk SSSR, Otd. Tekhn. Nauk, Metal. i Topliv., (5):62 (1961).

10. L. N. Aleksandrov, Izv. Vuz. Fiz.,No. (6):135 (1963).

11. B. G. Lazarev and O. N. Ovcharenko, Dokl. Akad. Nauk SSSR, 100 (5):875 (1955).

12. L. N. Aleksandrov, Uch. Zap. Mordovsk. Gos. Univ., No. 18, Ser. Fiz. -Mat. Nauk (Saransk, 1961).

13. L. N. Aleksandrov, Izv. Vuz. Fiz., (4): 102 (1961).

14. S. T. Konobeevskii, Zh. Eksp. i Teor. Fiz., 13 (6):185 (1943).

15. V. Gerold,' Aluminium (BRD), 37 (9):583 (1961).

16. V. M. Lomer, Collection: Vacancies and Other Point Defects in Metals and Alloys (Metallurgizdat, 1961).

17. B. Ya. Lyubov, Zh. Tekhn. Fiz., 20 (11):1314 (1950).

18. V. E. Neimark and I. B. Piletskaya, Probl. Metalloved. i Fiz. Metal., (3):376 (1952).

19. S. Z. Bokshtein, S. T. Kishkin, and L. M. Moroz, Issled. po zharoproch. Splavam. Akad. Nauk SSSR, (10):214 (1963).

PROCESSES OF THE REDISTRIBUTION AND HEALING
OF DEFECTS IN CRYSTALS ON HIGH-TEMPERATURE TREATMENT

É. N. Pogrebnoi

It is known that defects in crystal structure which develop during the solidification, heat treatment, and plastic deformation of metals and alloys may be redistributed and healed upon heating and high-temperature treatment [1-5]. The metallographical aspect of the redistribution and healing of defects has been inadequately investigated. The difficulties arising in metallographical study are to a great extent overcome if graphitizable silicon steels are used for the investigation [3, 4]. In these steels, graphite constitutes a phase which easily reveals disruptions in continuity. The use of graphitization as a means of "revealing" (decorating) damage is based on the fact that graphite forms much more rapidly in damaged parts of the matrix than elsewhere [3, 6, 7].

Examination of graphitizable steels [3, 4, 7] showed that the defects are redistributed and healed non-uniformly on high-temperature treatment. The grain boundaries play a large part in this. Coalescence, spheroidization, and growth of micropores may, depending on the temperature and length of homogenization, substantially change the number, size, shape, and distribution of the micropores and the degree of recovery of the matrix. Preferential precipitation and the coalescence of defects at grain boundaries on high-temperature (~1200°C) homogenization of steel led to the creation of a network of boundary pores, on the basis of which a graphite network developed during graphitization. In steel homogenized at lower temperatures this appeared rarely; more often the distribution of graphite inclusions forming in damaged parts of the matrix during graphitization remained almost uniform, and their number fell as the homogenization was prolonged.

We studied the redistribution and healing of defects on high-temperature homogenization in the austenite region for two cast-silicon steels, water-quenched from 1100°C to martensite:

	C	Si	Mn	P	S	Cr
A	1.35	0.72	0.34	0.038	0.028	0.02
B	0.92	1.34	0.37	0.040	0.033	0.03

The preliminary quench sharply increased the concentration of defects in the cast steel. On graphitization this appeared as a sharp rise in the number of centers and the rate of graphitization. In comparison with the original cast steel, the number of graphite inclusions forming on the defects increased 20 to 30 times, and they were distributed uniformly over the cross section.

Redistribution and healing of the defects before graphitization in the steel-A samples was effected by isothermal treatment in the austenitic state at 1050, 110, 1150, and 1200°C, and in the steel-B samples by special treatments in which various heating and cooling rates in the austenite region were combined with periods at fixed temperatures.* After such treatment, the samples were cooled in air and then briefly graphitized (10 h at 680°C) in order to "reveal" the defects in the matrix of the steels under examination by means of the graphite (using the method of [3, 4]). At the same time, the effect of the high-temperature treatment of quenched steel on its subsequent graphitization was determined.

Microscopic examination showed that the behavior of the quenched steel on graphitization depended considerably on the conditions of homogenization.

---

*The behavior of this steel during graphitization after isothermal homogenization at 900−1200°C was studied in [3, 4].

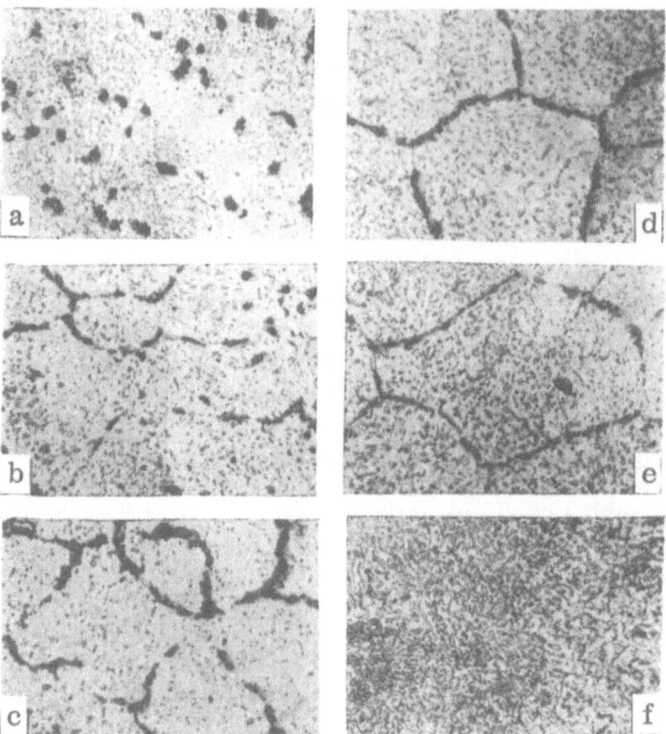

Fig. 1. Structural variations in homogenized steel A after graphization for 10 h at 680°. After quenching, the samples were homogenized at 1200° for a) 1; b) 5; c) 15; d) 90; e) 120; f) 180 min. Magnification × 500.

In the samples of steel A which underwent graphization after isothermal homogenization at 1050, 1100, 1150, and 1200°C, it was observed that,depending on the length of homogenization, the graphite network first appeared and then disappeared, as noted in [3, 4, 6]. The nature of these changes is shown in Fig. 1. In steel samples maintained at 1200° for 1 min, the graphite inclusions were distributed almost uniformly at graphization (Fig. 1a). Increasing the period to 5 min resulted in a nonuniform distribution of graphite inclusions, and parts of a disjointed graphite network appeared in the structure (Fig. 1b). On annealing samples homogenized for a longer time (15 min), the graphite network forming on the defects reached its maximum development (Fig. 1c); then, with a prolongation of the period of homogenization to 120 min, the graphite network gradually vanished, and was replaced by a network of secondary cementite (Fig. 1d, e). In samples homogenized for long periods (more than 120 min), no graphite network developed on annealing, and only a few graphite inclusions appeared in the structure; in steel homogenized for 180 min at 1200°, none at all appeared after annealing for 10 h at 680° (Fig. 1f).

Similar structural changes on graphitization were also observed in the samples of quenched steel A homogenized at lower temperatures (1050, 1100, and 1150°C). The time required for the graphite network 1) to begin to appear, 2) to complete its development, and 3) to vanish entirely in homogenized samples of quenched steel during graphitization varied considerably as the homogenization temperature was varied (Fig. 2). The graphite network

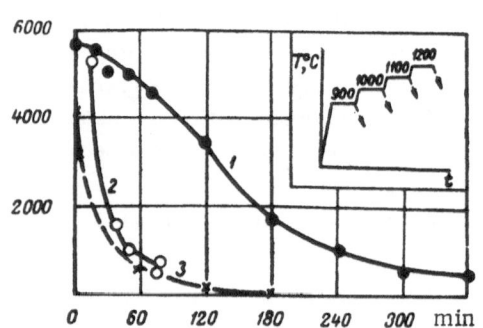

Fig. 2. Variation of the parameters defining the homogenization process with homogenization temperature.

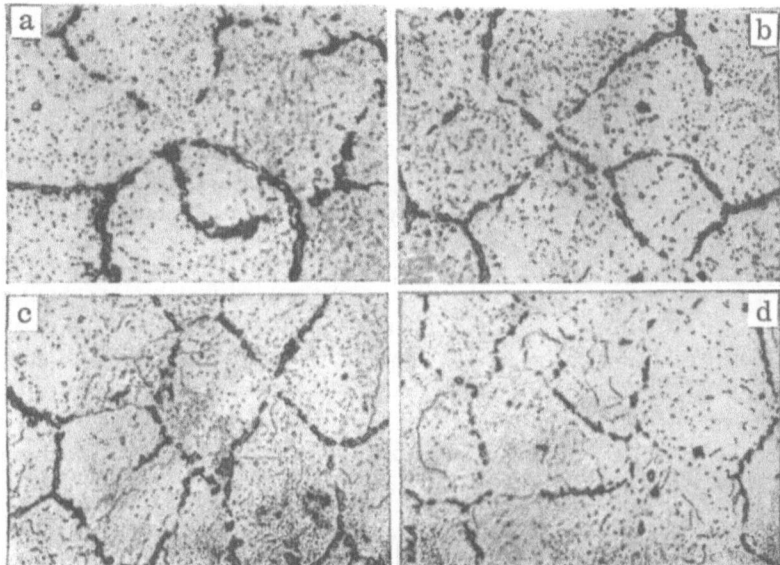

Fig. 3. Graphite network formed on graphitization of homogenized steel A. Homogenization conditions: a) 15 min at 1200°, b) 30 min at 1150°, c) 45 min at 110°, d) 90 min at 1050°; magnification × 500.

appeared and vanished with increasing rapidity with increases both in the rates of annealing of homogenized steel A, and in the homogenization temperature (Fig. 3).

In isothermally homogenized samples of steel A, the tendency to form a graphite network during graphitization and its duration time are more noticeable than in similar samples of steel B, where the graphite network only appeared stable on annealing after homogenization at 1200°, and for lower temperatures only appeared occasionally.

Etching with sodium picrate showed that the graphite network formed during graphitization on the boundary pores coincided with both the primary and new polygonization boundaries arising as a result of the grouping of dislocations and the settling of excess vacancies on these.

As the temperature and duration of homogenization of the quenched steel increase, the austenite grain is perfected and greatly enlarged; the subsequent subcritical graphitization of cementite in the steel is retarded. In samples protractedly homogenized in the austenite state, when quenching and shrinkage microdiscontinuities have healed, after a 10-h anneal at 680° the cementite does not graphitize at all (Fig. 2, line 4).

The possibility of forming a graphite network on graphitization of homogenized steel was also examined for samples of quenched steel B submitted to step-by-step homogenization in the austenite region. In samples of this steel, after homogenization at 900 and 1200°, the number of graphite inclusions formed on graphitization in the defective matrix declines as the duration of homogenization increases (Fig. 4, curves 1, 3); their distribution in the matrix of the steel homogenized at 900° remains uniform, and their size gradually increases, but in steel homogenized at 1200° it changes nonuniformly, as a network [3, 4].

If the samples of quenched steel B are homogenized step-by-step with 15 min periods at 900, 1000, 1100, and 1200° (Fig. 4) in order to accelerate the processes of redistribution and healing of defects, then the appearance of the subsequent graphitization shows that the number of graphite inclusions forming on the defects falls markedly with increasing homogenization temperature (Fig. 4, curve 2). This is also clearly seen in the structure of the samples in question (Fig. 5). It is important to note that the graphite network only develops in samples homogenized on annealing at 1200° (Fig. 5d). Each new rise in temperature led to an intensification

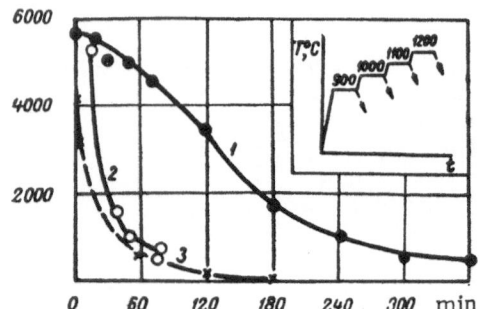

Fig. 4. Variation in the number of graphite inclusions on graphitization of quenched steel B, which undergoes homogenization after quenching 1) at 900°, 2) in the manner indicated in the top right corner, 3) at 1200°.

of the redistribution and healing of defects, with no change in the laws of their flow, until these were disrupted near 1200°, when a network of boundary slot pores developed in the structure, this being connected with an increase in the part played by structural elements (primary and polygonized boundaries, boundary and surface diffusion) and recrystallization in the austenite.

The formation of a graphite network on annealing was also observed in samples submitted to step-by-step homogenization, in which samples of quenched steel B were first homogenized for 15 or 60 min at 900° (to obtain larger and more equiaxial micropores), and then again at 1200° for periods of 5, 15, 30, 60, and 120 min, after which they were cooled in air and graphitized at 680° (Fig. 6).

Also investigated were the processes of redistribution and healing of defects in steel in which a network of boundary pores was first formed by high-temperature homogenization and then removed by low-temperature homogenization. For this purpose, samples of quenched steel B were homogenized for 15 min at 1200° and then submitted to homogenization at 940° for 15, 30, 60, and 120 min. On annealing, the samples homogenized at 1200° developed a graphite network (Fig. 7a), while in those homogenized additionally at 940° the pores gradually healed, considerably reducing the tendency to form a graphite network on annealing (Fig. 7b, c); instead a chain-like arrangement of graphite inclusions developed (Fig. 7d), though this was not observed in samples homogenized at 940° for a long period.

We may conclude from the results presented that crystal-structural defects existing in case and quenched steel are only partly healed on rapid heating. In order to remove them, additional high-temperature treatment

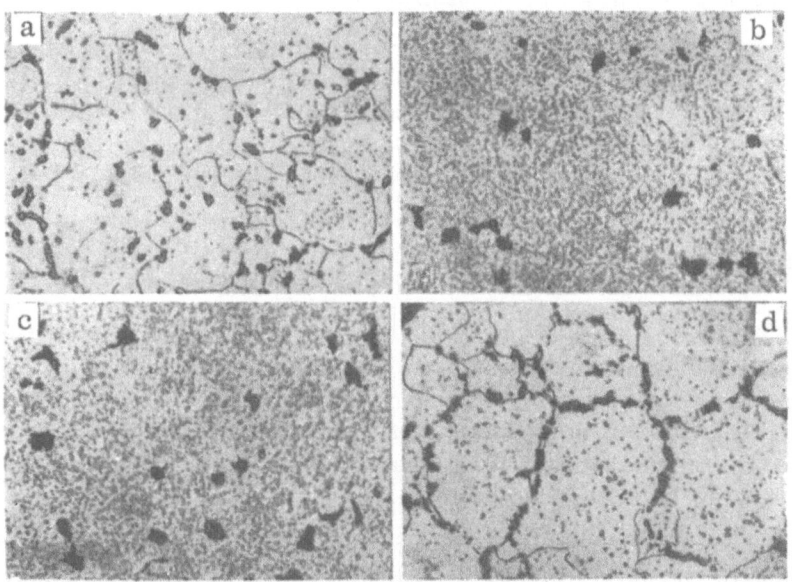

Fig. 5. Structural changes on graphitization of quenched steel B submitted to step-by-step homogenization after quenching: a) 15 min at 900°, b) 15 min each at 900 and 1000°, c) 15 min each at 900, 1000, and 1100°, d) 15 min each at 900, 1000, 1100, and 1200°. Magnification × 500.

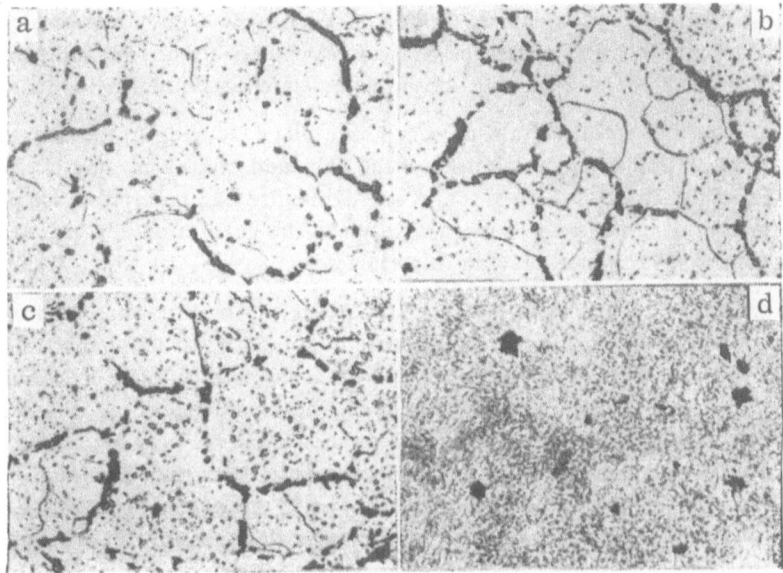

Fig. 6. Structural changes on graphitization in step-by-step homogenized steel B ($900 \rightarrow 1200°$). Periods at $1200°$: a) 5, b) 15, c) 60, d) 120 min. Magnification × 500.

is required. The molecular picture of the processes of redistribution and healing of defects in damaged crystals is apparently as follows.

In cast steel quenched from high temperatures, the super-equilibrium defect concentration is considerably higher than in unquenched steel. During solidification, crystallization defects form in the steel in the shape of single and grouped vacancies, dislocations, shrinkage, and diffusion micropores. The density of the dislocations in crystals obtained by dendrite crystallization may reach $\sim 10^6 - 10^8$ lines/cm^2 [8]. After quenching to martensite the crystals of the solid solution are supersaturated with carbon and vacancies, the block and grain structure

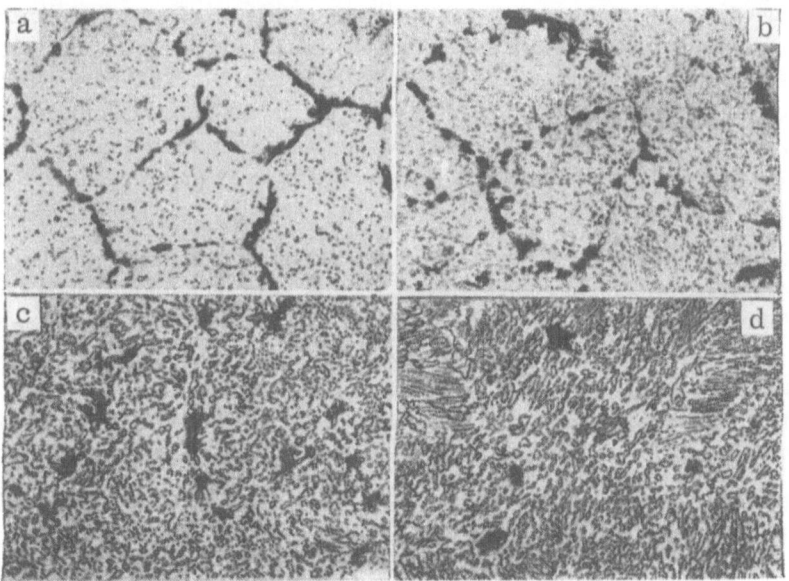

Fig. 7. Structural changes on graphitization in step-by-step homogenized steel B ($1200 \rightarrow 940°$). Magnification × 500.

is severely broken up, and the dislocation density in the crystals rises to $\sim 10^{12}$ lines/cm^2 [9] and considerably increases the number of microdiscontinuities in the form of accumulations of vacancies, pores, and microcracks. In the matrix of quenched steel, these super-equilibrium defects are distributed more evenly than in cast steel.

On the heating of quenched steel, certain processes take place, leading to the decomposition of the martensite and to the gradual recovery of the lattice of the damaged crystals by way of the interaction, redistribution, and healing of the defects. Both graphite and cementite can precipitate from the carbon-supersaturated solid solution, but cementite does so more rapidly. In the generation and growth of the high-carbon phases, the defects of the crystal structure play an important part.

On rapid heating, the damaged crystals recover only partially, mainly as a result of the interaction and displacement of vacancies and dislocations possessing fairly high mobility. The main sinks for excess vacancies are dislocations, grain boundaries, and sub-boundaries, as well as the free surfaces of discontinuities, new phases. and the sample itself. Settling predominantly on grain boundaries and dislocations, the excess vacancies may form pores [2, 5, 10]. According to [10], pores formed around dislocations may be as large as 10^{-6} cm in size, and arise when the vacancies approach the dislocations more rapidly than they disappear at the steps. The condensation of vacancies in pores at grain boundaries and even in the centers of dislocations is accompanied by a gain in energy; this can take place at dislocations for vacancy concentrations below 10^{-5} at room and lower temperatures [10].

With increasing temperature, the mobility of the vacancies and dislocations also increases, and the discharge of vacancies at dislocations and other imperfections accelerates. This intensifies not only the precipitation, coalescence, and spheroidization of the carbide phase, but also the relaxation of stresses, polygonization, and recrystallization in quenched steel. Hence the perfection of the lattice in the damaged crystals is accelerated, and the concentration of vacancies and dislocations in them diminishes.

The redistribution of silicon between the carbide phase and solid solution and in the inhomogeneous solid solution taking place on heating to the temperature of the austenitic state, together with the phase changes associated with the polymorphic transformation and dissolution of cementite, lead to a change in the dislocation and grain structure and the generation of new vacancies and dislocations. On rapid heating these are also partly redistributed and healed.

The coarser and less mobile defects retained during rapid heating up to the austenitic temperature range are removed by subsequent maintenance in the austenite state. The rate of healing is determined by the temperature and duration of maintenance in the austenitic state, the conditions of heating and cooling in this region, and the processes of polygonization and recrystallization in the austenite. The redistribution and healing of defects in a polycrystalline austenite matrix, just as in sintering powder compositions, is apparently effected by way of the displacement of vacancies and dislocations, predominantly their emergence at block and grain boundaries and pore surfaces; in this an important part is played by coalescence, spheroidization, and the growth of micropores [2-5].

On low-temperature homogenization, these processes take place comparatively slowly. As they coalesce, the micropores become coarser and their number diminishes, and they spheroidize themselves as well into equiaxial pores, not infrequently with clearly expressed faces. The role of spheroidization in changing the form of pores is also clearly revealed in the samples in which a network of boundary slot pores was formed during high-temperature homogenization and which were then studied under low-temperature homogenization. From the variation in the shape of the graphite inclusions formed on these pores during graphitization, we may conclude that the slot pores can break down and transform into equiaxial pores (Fig. 7).

On increasing the homogenization temperature, the processes of redistribution and healing of defects accelerate, and at the same time the role of structural defects (primary and polygonization boundaries, boundary and surface diffusion) increases, as does that of recrystallization, granulation, and homogenization in the austenite. As they coalesce, the boundary pores expand more rapidly than the intragrain pores. This appears not only in the change in their number and magnitude but also in the form and character of the distribution.

For high-temperature homogenization, the boundary pores expand along the primary and polygonization boundaries, forming a network of slot pores. These changes are also found in steels in which the micropores were specially grown to coarser dimensions during low-temperature homogenization and had a compact form. On high-temperature homogenization these also coalesce, with the formation of a boundary network of pores, on which a graphite network forms during subsequent graphitization (Figs. 5 and 6).

It may therefore be seen that at certain stages of homogenization the micropores can coalesce faster than they can spheroidize. In subsequeng holding, when the coalescence of the pores is retarded so much that the rates of spheroidization and pore growth begin to prevail, the boundary slot pores break down into finer ones and transform into equiaxial pores.

From the similarity between the structural pictures of the appearance and disappearance of the graphite network during the annealing of homogenized steels A and B (which differ in carbon and silicon content and original structure), we may conclude that the character of the redistribution and healing of defects in these steels on homogenization in the austenitic state remains as before, the process resulting from diffusion and dislocation acts. Since the duration of the existence of boundary pores on homogenization is considerably greater in the steel with the higher carbon content (from the tendency to form a graphite network on subcritical annealing), we may suppose that in this steel, apart from defects of shrinkage, diffusion, and quenching origin, and defects formed with austenitization and with the decomposition of the martensite, an important part in the development of porosity is also played by defects arising in the matrix as a result of the dissolution of secondary cementite, situated in the structure predominantly along grain boundaries. After such dissolution, the concentration of vacancies and dislocations on these grain boundaries evidently rises substantially, and this also accelerates the appearance and coalescence of boundary micropores on homogenization.

The processes of redistribution and healing of defects play an important part in the recovery of damaged crystals. An increase in the degree of recovery of the matrix and a decrease in the number of micropores and other defects in cast and quenched steels greatly impede their subsequent graphitization. The more completely the defects are removed from the steel, the more strongly will graphite formation be retarded.

Literature Cited

1. F. F. Vol'kenshtein, Statistical Phenomena in Heterogeneous Systems, Izd. Akad. Nauk SSSR (Moscow— Leningrad, 1949).
2. Ya. E. Geguzin, Usp. Fiz. Nauk (2):61 (1957).
3. K. P. Bunin and É. N. Pogrebnoi, Izv. Akad. Nauk SSSR, Otd. Tekhn. Nauk (12): (1955).
4. É. N. Pogrebnoi, Izv. Akad. Nauk SSSR, Otd. Tekhn. Nauk (1): (1961).
5. I. M. Fedorchenko and L. A. Andrievskii, Fundamentals of Powder Metallurgy (Izd. Akad. Nauk Ukr. SSR, 1961).
6. K. P. Bunin and N. M. Danil'chenko, Dokl. Akad. Nauk SSSR 72 (5):(1950); 82 (3):(1952).
7. K. P. Bunin, A. A. Baranov, and E. N. Pogrebnoi, Graphitization of Steel (Izd. Akad. Nauk Ukr. SSR, 1961).
8. W. J. Tiller, Appl. Phys. 26:611—18 (1958).
9. J. Kelly, and I. J. Nutting, Iron Steel Inst., (London) 19 XII:(1960).
10. P. Coulon and J. Friedel, Collection: Dislocations and the Mechanical Properties of Crystals [Russian translation], IL, 1960.

THEORY OF THE DECOMPOSITION OF A FERROMAGNETIC ALLOY

V. M. Danilenko

It was shown in [1, 2] that the processes of ordering and magnetization in an alloy made up of ferromagnetic (or antiferromagnetic) components are mutually interrelated. One parameter of this interrelation is the coefficient for the part of the exchange energy of the magnetic electrons depending on the degree of long-range order. This is proportional to the following combination of exchange integrals for the pairs of atoms A-B, A-A, and B-B:

$$\alpha - 2A_{AB} - A_{AA} - A_{BB}. \tag{1}$$

For $\alpha > 0$ the ordering of the atoms causes ferromagnetism to appear in the alloy, and, conversely, magnetization produces ordering. For $\alpha < 0$ ordering results in the establishment of antiferromagnetic ordering of the spins of the magnetic electrons. Ferromagnetic or antiferromagnetic ordering of the spins is decided by the sign of the mean exchange integral in the alloy, which for an alloy of the β-brass type equals

$$\overline{A} = A_0 + \alpha x^2, \tag{2}$$

where

$$A_0 = aA_{AA} + bA_{BB} + ab\alpha. \tag{3}$$

Here a and b are the concentrations of atoms A and B in the alloy (a + b = 1), and x is the degree of long-range order, determined by the formula

$$x = p_A^1 - a, \tag{4}$$

where p_A^1 is the probability that atom A replaces a node of the first kind.

For $\overline{A} > 0$ the alloy is ferromagnetic, and for $\overline{A} < 0$ it is antiferromagnetic.

The order in the distribution of atoms is determined for the same alloy by the quantity

$$\overline{w} = w \pm \frac{\alpha}{2} y^2, \tag{5}$$

where w is the energy of ordering and y is the magnetization parameter;

$$y = \frac{2r - \dfrac{N}{2}}{\dfrac{N}{2}}, \tag{6}$$

where r is the number of "right-hand" spins in one of the sublattices of the alloy.

135

In formula (5) the plus sign stands for ferromagnetism, while the minus sign signifies antiferromagnetism.

For certain values of α, in fact, on satisfying the conditions

$$|\alpha| > 2w,$$
$$\pm\,\alpha < 0,$$

(6a)

the quantity $\overline{w}$ becomes negative, and ordering in such alloys becomes impossible.

As we know, the case $w < 0$ corresponds to the decomposition of a nonferromagnetic alloy into two dis-ordered solid solutions. Clearly, even for a ferromagnetic alloy, we must take account of this possibility. In this paper we present a statistical calculation of the phase diagram of a decomposing ferromagnetic alloy with body-centered cubic lattice, based upon the same assumptions as in [1, 2]. The calculation is carried out in the Gorskii—Bragg—Williams approximation, allowing for interaction in the first coordination sphere. For the exchange energy of the magnetic electrons Heisenberg's formula is used.

The free energy ψ of a homogeneous solid solution in this approximation is expressed by the formula

$$\psi = -N\,\frac{z}{2}\left[aV_{AA} + bV_{BB} + abw + \frac{1}{2}\,A_0 y^2\right] +$$
$$+ NkT\left[a\ln a + b\ln b + \frac{1+y}{2}\ln\frac{1+y}{2} + \frac{1-y}{2}\ln\frac{1-y}{2}\right].$$

(7)

Here N is the number of atoms in the alloy, z is the coordination number (in our case z = 8), V_{AA} and V_{BB} are the interaction energies of atom pairs A—A and B—B without allowing for exchange energy, taken with inverse sign.

In calculating for one atom of the alloy, the free energy will equal

$$\varphi = \frac{\psi}{N}\,.$$

(8)

In a state of thermodynamic equilibrium, the following conditions must be satisfied for an alloy consisting of two solid solutions of concentrations a_1 and a_2:

$$\frac{\partial\varphi_1}{\partial y_1} = 0,\quad \frac{\partial\varphi_2}{\partial y_2} = 0,\quad \frac{\partial\varphi_1}{\partial a_1} = \frac{\partial\varphi_2}{\partial a_2}\,,$$

$$\varphi_1 - \varphi_2 = \frac{1}{2}\left(\frac{\partial\varphi_1}{\partial a_1} + \frac{\partial\varphi_2}{\partial a_2}\right)(a_1 - a_2),$$

(9)

which may be reduced to the form

$$\ln\frac{1+y_1}{1-y_1} - \frac{z}{kT}\,A_{01}y_1 = 0;$$

(10)

$$\ln\frac{1+y_2}{1-y_2} - \frac{z}{kT}\,A_{02}y_2 = 0;$$

(11)

$$\ln\left(\frac{a_1}{b_1}:\frac{a_2}{b_2}\right) - \frac{z}{2kT}\left[\frac{1}{2}(A_{AA}-A_{BB})(y_1^2-y_2^2) - \right.$$

$$\left. - 2w(a_1-a_2) + \frac{1}{2}\alpha(y_1^2-y_2^2) - \alpha(a_1y_1^2-a_2y_2^2)\right] = 0; \tag{12}$$

$$a_1\ln a_1 + b_1\ln b_1 - a_2\ln a_2 - b_2\ln b_2 -$$

$$-\frac{1}{2}\ln\left(\frac{a_1a_2}{b_1b_2}\right)(a_1-a_2) + \frac{1+y_1}{2}\ln\frac{1+y_1}{2} +$$

$$+\frac{1-y_1}{2}\ln\frac{1-y_1}{2} - \frac{1+y_2}{2}\ln\frac{1+y_2}{2} - \tag{13}$$

$$-\frac{1-y_2}{2}\ln\frac{1-y_2}{2} - \frac{z}{8kT}(y_1^2-y_2^2)[A_{01}+A_{02}+$$

$$+\alpha(a_1-a_2)^2] = 0.$$

In formulas (9) to (13), y_1 and y_2 and A_{01} and A_{02} relate to the phases with concentration a_1 and a_2.

Equations (10) and (11) describe the well-known temperature variation of spontaneous magnetization in an alloy of constant composition. However, in view of the fact that the concentration changes on decomposition of the alloy, this relationship has a complex character in the actual alloy. The concentration of the phases during decomposition [determined by formulas (12) and (13)] also depends on the magnetization of the two phases (y_1 and y_2). In the general case, the configuration of the two-phase region has a complicated form, and numerical solution of equation system (10) to (13) is required for its determination. In certain cases, however, an analytical solution may be obtained.

1. If $\alpha = 0$ and $A_{AA} = A_{BB} = A$, then $A_{01} = A_{02} = A$, and the system of equations takes the simple form

$$\ln\frac{1+y}{1-y} - \frac{z}{kT}Ay = 0;$$

$$\ln\left(\frac{a_1}{b_1}:\frac{a_2}{b_2}\right) + \frac{zw}{kT}(a_1-a_2) = 0; \tag{14}$$

$$a_1\ln a_1 + b_1\ln b_1 - a_2\ln a_2 - b_2\ln b_2 - \frac{1}{2}\ln\left(\frac{a_1a_2}{b_1b_2}\right)(a_1-a_2) = 0.$$

Here we have considered that for $A_{01} = A_{02}, y_1 = y_2 = y$, and hence all the alloys have the same magnetic properties, described by the first of equations (14). Substituting for (a_1-a_2) from the second expression into the third, we obtain the equation

$$a_1\ln a_1 + b_1\ln b_1 - a_2\ln a_2 - b_2\ln b_2 -$$

$$-\frac{1}{2}\frac{kT}{zw}\left\{\left(\ln\frac{a_2}{b_2}\right)^2 - \left(\ln\frac{a_1}{b_1}\right)^2\right\} = 0, \tag{15}$$

which has two solutions:

$$a_1 = a_2 = a, \tag{16}$$

$$a_1 = b_2, \quad b_1 = a_2. \tag{17}$$

The first solution corresponds to the homogeneous state of the solid solution and is unstable in the two-phase region; the second corresponds to decomposition into phases of "symmetric" concentration.

The decomposition curve is determined by the second of equations (14):

$$\ln \frac{a_1}{b_1} + \frac{zw}{kT}\left(a_1 - \frac{1}{2}\right) = 0, \tag{18}$$

i.e., it has a well-known form [4]. As we see, in this case magnetization and decomposition of the alloy take place independently of each other.

2. If $A_{AA} = A_{BB} = A$, but $\alpha \neq 0$, the equations become more complicated, but even in this case solutions (16) and (17) hold, and the decomposition curve is now determined by the system of equations

$$\ln \frac{1+y}{1-y} - \frac{z}{kT}(A + \alpha a_1 b_1) y = 0;$$
$$\ln \frac{a_1}{b_1} + \frac{z}{kT}\left(w + \frac{\alpha y^2}{2}\right)\left(a_1 - \frac{1}{2}\right) = 0. \tag{19}$$

This system of equations is analogous to that defining the mutual connection between the processes of magnetization and ordering [1, 2], which is especially noticeable on introducing the variable

$$c = 2a_1 - 1. \tag{20}$$

For equation system (19) we obtain

$$\lambda = \ln \frac{1+y}{1-y} - t\left(B - \frac{\alpha}{4}c^2\right)y = 0;$$
$$\mu = \ln \frac{1+c}{1-c} - t\left(\frac{|w|}{2} - \frac{\alpha}{4}y^2\right)c = 0, \tag{21}$$

since w < 0 for decomposing alloys. In (21) we have used the notation

$$B = A + \frac{\alpha}{4}, \quad t = \frac{z}{kT}. \tag{22}$$

These equations may be studied by the method used in [1, 2]. The conditions $\partial\lambda/\partial y = 0$ and $\partial\mu/\partial c = 0$ correspond to singular points on the curve relating y and c to T.

In the first case this is the Curie point for the phase of composition $a_1 = (1 + c)/2$:

$$t^{mag} = \frac{2}{B - \frac{\alpha}{4} c^2} \qquad (23)$$

$$T^{mag} = \frac{z}{2k} \left(B - \frac{\alpha}{4} c^2 \right) = \frac{z}{2k} (A + \alpha a_1 b_1). \qquad (24)$$

In the second case this is the critical point of the decomposition, defined by the formula

$$t^{decomp} = \frac{4}{|w| - \frac{\alpha}{2} y^2} \qquad (25)$$

or

$$T^{decomp} = \frac{z}{4k} \left(|w| - \frac{\alpha}{2} y^2 \right). \qquad (26)$$

These formulas are analogous to those obtained for the ordering temperatures and Curie points in [1, 2]. The case of the decomposition of the alloy supplements the cases considered in these papers. Just as antiferromagnetism and ferromagnetism transform into each other with change of the sign of the quantity $\overline{A}(2)$, so also does decomposition correspond to the ordering of the alloy with the change of the sign of the quantity $\overline{w}(5)$. It is easy to generalize our equation (26) to the case of the antiferromagnetic alloy; no change need be made other than the altering the sign in front of the term $(\alpha/2)y_2$.

Analogously with the work of [1, 2], we may also study here the behavior of the "equilibrium curve" (relationship between y, c, and T) near its singular points. In view of the total symmetry of the processes in question, the results obtained in [1] may be entirely transferred to the decomposing alloy (changing α into $-\alpha$). Hence both the magnetic transformation curve and the decomposition curve can have an anomalous form (corresponding to a transformation of the first kind in the ordered state). Here the critical point defined by equation (26) lies below the actual upper boundary of the two-phase region. This takes place for large absolute values of α and for $\alpha < 0$ for the ferromagnetic alloy and for $\alpha > 0$ for the antiferromagnetic alloy. For the opposite sign of α there may occur a decline in the parameter y on decomposition and reversal of the decomposition curve on magnetization of the alloy. Finally, for a large absolute value and opposite sign of α, ferromagnetism changes into antiferromagnetism, and decomposition into ordering.

3. If $\alpha = 0$, but $A_{AA} \neq A_{BB}$, solution (17) does not hold, since now $A_{01} \neq A_{02}$ and $y_1 \neq y_2$ even for $\alpha_1 = b_2$. In this case the two-phase region ceases to be symmetrical, and cannot be expressed in a simple analytical fashion.

If the difference $A_{AA} - A_{BB} = \Delta$ is small there may be a displacement of the critical point relative to its position in the nonmagnetized alloy:

$$\frac{\Delta T^{decomp}}{T} \approx \frac{y^2(1 - y^2)}{1 + \frac{2A}{w}(1 - y^2)} \left(\frac{\Delta}{w} \right)^2; \qquad (27)$$

V. M. DANILENKO

$$\Delta a_{\mathrm{cr}} \approx \frac{\frac{3}{4} y^2 (1 - y^2) \left[1 - \frac{5}{3} y^2 + 2 \frac{A}{w} (1 - y^2)^2 \right]}{\left[1 + 2 \frac{A}{w} (1 - y^2) \right]^3} \left(\frac{\Delta}{w} \right)^3, \qquad (28)$$

where

$$A = \frac{A_{AA} + A_{BB}}{2} .$$

Literature Cited

1. V. M. Danilenko and A. A. Smirnov, Fiz. Metal. i Metalloved., 17:337 (1962).
2. V. M. Danilenko, D. R. Rizdvyanetskii, and A. A. Smirnov, Ukr. Fiz. Zh., 8:294 (1963).
3. Ya. S. Umanskii, B. N. Finkel'shtein, M. E. Blanter, S. T. Kishkin, N. S. Fastov, and S. S. Gorelik, Physical Metallurgy (Metallurgizdat, 1955).

STUDY OF THE INITIAL STAGES OF CRYSTALLIZATION
OF SPHEROIDAL GRAPHITE IN Ni−C ALLOY

I. E. Bolotov

The crystallization of many substances takes place predominantly by the formation of spherulites. Graphite may crystallize in this manner on solidification of cast iron, Ni−C, and Co−C alloys [1], as well as certain other metallic alloys including carbon; so likewise may many inorganic substances (S, Se, etc.), while spherulite formation is the normal manner of crystallization in polymers [2].

A characteristic feature of a spherulite is the regular orientation of the crystal lattice within it. Thus, in graphite, the basal plane of its crystal lattice is perpendicular to the radius of the spherulite [1]. In polymer spherulites, the crystal lattice is oriented so that the direction of the molecules is perpendicular to the radius [2].

The aim of the present paper is to study the crystallization of graphite spherulites in the early stages, in which they still have submicroscopic dimensions (i.e., cannot be examined under the optical microscope). For this it was necessary to interrupt the growth of the crystals at an early stage of crystallization. Attempts to produce small spherulites suitable for examination in the electron microscope on solidification of Ni−C alloy have as yet not yielded positive results, despite the use of rapid cooling. We therefore studied the manner of precipitation of graphite crystals upon the decomposition of supersaturated Ni−C solid solution (on aging). By varying the aging temperature, it was possible to vary at will the dimensions of the precipitating graphite crystals and to study these in the early stages of formation under the electron microscope.

Conditions of Spherulite Formation on Aging Ni − C Alloy

Alloys were prepared from electrolytic 000-type Ni; these contained 0.65% carbon, corresponding to the maximum solubility of carbon in nickel. The alloys were quenched from 1200°C immediately after solidification. By aging the alloy at 650°C for 1 h, it was possible to obtain graphite crystals several microns in size, which could be studied under the optical microscope. It was found that on annealing the alloy in sealed quartz ampoules the graphite crystallized in the form of round crystals, showing the radial structure (Maltese cross) characteristic of spherulites in polarized light (Fig. 1). On annealing the alloy without a protective atmosphere (in air), the graphite crystals were of irregular form and failed to exhibit a radial structure in polarized light (Fig. 2). It may be assumed that the crystallization of the graphite is affected by hydrogen contained in the nickel (on annealing in sealed ampoules this could not be completely removed from the alloy; on annealing in air it could). In order to verify this assumption we carried out annealing at the same temperature (650°C) in vacuum and in a hydrogen atmosphere. In the first case the graphite was of irregular form and lacked radial structure. In the second it crystallized as regular spherulites which had radial structure. We may thus conclude that aging Ni−C alloy in the presence of hydrogen impurities leads to the precipitation of graphite spherulites. When an alloy devoid of hydrogen is aged, however, graphite crystals of irregular form precipitate.

Structure of Spherulites in the Initial Stages of Growth

The initial stages in the precipitation of both spheroidal and irregular graphite crystals were studied under the electron microscope.

After metallographic sections were prepared from samples of alloys aged at various temperatures, these were etched in 1% solution of bromine in methyl alcohol. As a result of the dissolution of the metallic matrix,

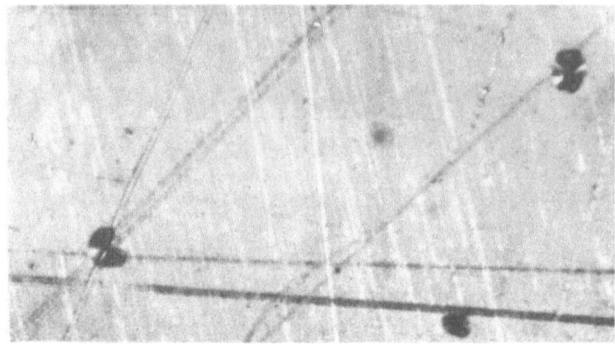

Fig. 1. Optical micrograph of Ni–C alloy after annealing
in sealed ampoules at 650° for 1 h. Magnification ×1200.
Polarized light.

the graphite crystals projected above its surface. A carbon or lacquer replica was laid on the etched surface.
After removal of the replica, some of the graphite crystals were taken up by the replica, while others remained
in the metal matrix, leaving the imprint of their surface on the replica. Thus both the graphite crystals themselves and the depressions left by them in the replica, giving the surface of the crystals, were studied in the
electron microscope. In both cases photographs were taken in a stereographic holder at two different angles.
The photographs were examined in a stereoscope, enabling conclusions to be drawn as to the form of the crystals.

Figure 3 shows electron micrographs of typical graphite spherulites observed after aging at 500 to 550°C for
1 h in sealed quartz ampoules. The most typical feature is the sharply expressed hexagonal form of the crystals.
Some of the crystals are transparent to 40–50 keV electrons, others are semitransparent, and a few are opaque.

Figure 4 shows electron micrographs of imprints left by graphite spherulites in a carbon replica. The results
of examining stereophotographs of the actual crystals and the traces from their surface may be summarized in
the scheme of Fig. 5, which illustrates typical crystal forms. Those most frequently encountered include:
hexagonal prisms (Figs. 4a and 5a), hexagonal pyramids (Figs. 4b and 5b), tubes coiled in the shape of hexagonal
toroids (Figs. 4c and 5c), and tubes coiled in the shape of hexagonal helicoids (Figs. 4d and 5d).

The radial structure of the graphite crystals observed in polarized light (see Fig. 1) indicates that each
graphite inclusion is a polycrystalline spherulite rather than a single crystal. Radial structure is found after
annealing the very smallest crystallites still observable under the optical microscope (crystals around 1 μ in
size) in the presence of hydrogen. It is therefore natural to expect that crystals studied under the electron

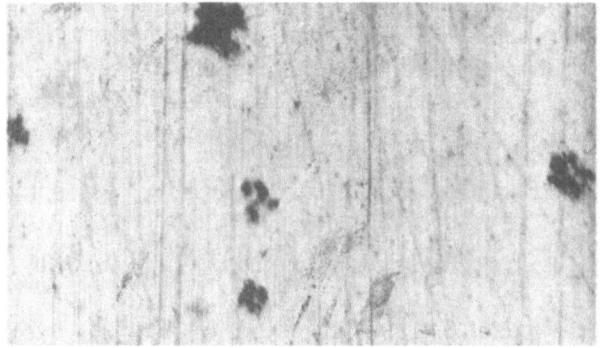

Fig. 2. Optical micrograph of Ni–C alloy after annealing
in air at 650° for 1 h. Polarized light.

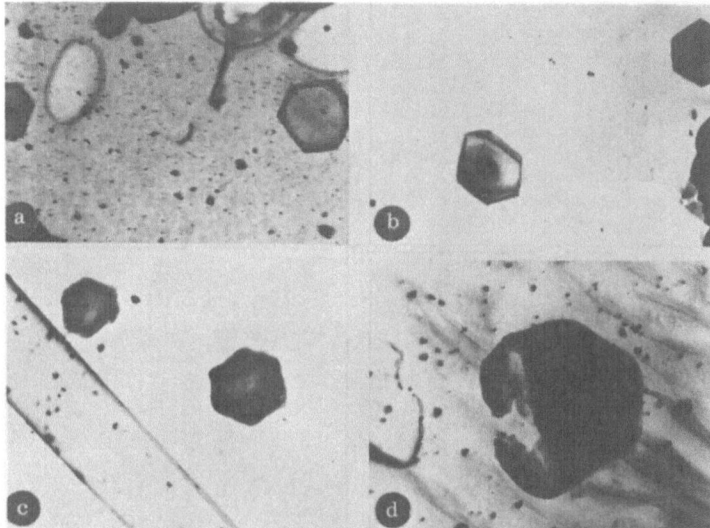

Fig. 3. Electron micrographs of graphite crystals after annealing in sealed ampoules at 550° for 1 h. Lacquer replica. Magnification × 40,000.

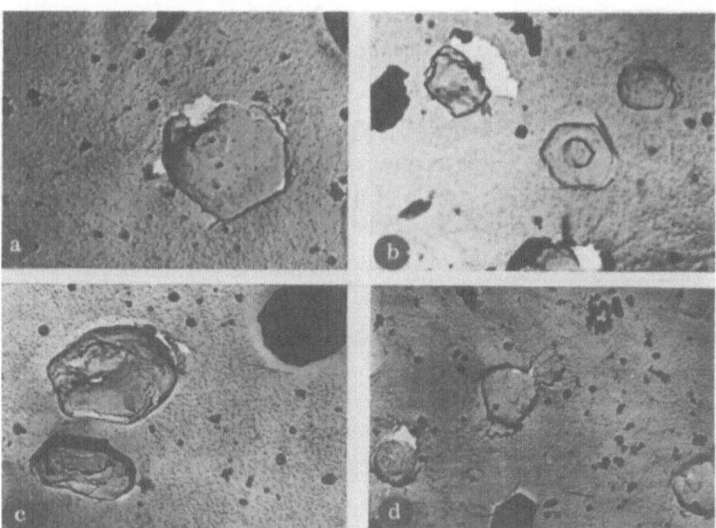

Fig. 4. Electron micrographs of Ni—C alloy after annealing in sealed ampoules at 550° for 1 h. Carbon replica. Magnification × 40,000.

microscope (0.1 to 1 μ in size) will also prove to be polycrystalline. This was confirmed by an electron micro-diffraction study of individual crystals.

The electron diffraction photographs obtained from individual crystals (Fig. 6) constitute a cross section of the reciprocal lattice interference sphere of graphite. They differ, however, from the picture which would be expected from a single crystal. There appear maxima, formed as a result of points with nonzero l falling on the plane of the diffraction sphere, owing to the existence of disorientation around an axis perpendicular to the ray. Beside this, there is an azimuthal disorientation, indicated by the extension of the spots around

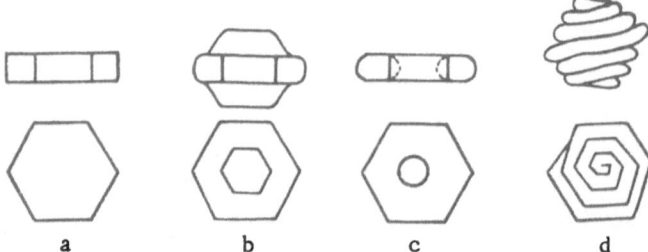

Fig. 5. Types of graphite crystals in Ni—C alloy annealed in sealed ampoules.

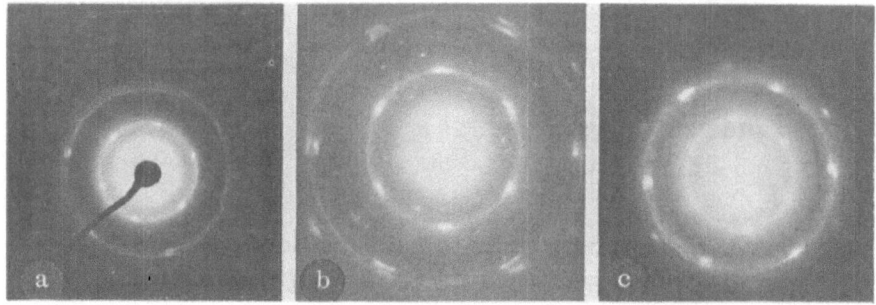

Fig. 6. Electron diffraction photographs of graphite crystals. a) Crystal
shown in Fig. 3b: b) crystal shown in Fig. 3a; c) crystal shown in Fig. 3c.

the circle. Evidently the external surface of the crystal is the basal plane of the graphite lattice. The existence of azimuthal disorientation may be explained by the fact that, in forming the hexagonal tube, neighboring parts of the tube are turned one against the other.

From a study of electron microscopic photographs (see Figs. 3 and 4), we may conclude that, within a single grain of the metal matrix, there is a single orientation of crystals. The precipitating crystals are oriented regularly with respect to the crystal lattice of the metal. On passing to another grain, there is a change in the orientation of the hexagonal crystals. The same conclusion may be reached on the basis of the optical micrographs, since in some cases the hexagonal form of the crystals can be observed in the optical microscope (see Fig. 1).

After annealing samples at 500—550° in air (i.e., in conditions ensuring the removal of the hydrogen), the crystals shown in Fig. 7 precipitate. The hexagonal shape is noticeable in the finer crystals. These are not, however, prisms with relatively level faces; the surface of the prisms is covered by a large number of unsystematically disposed mounds. In later stages of growth, graphite of flocculent form is obtained (Fig. 7b).

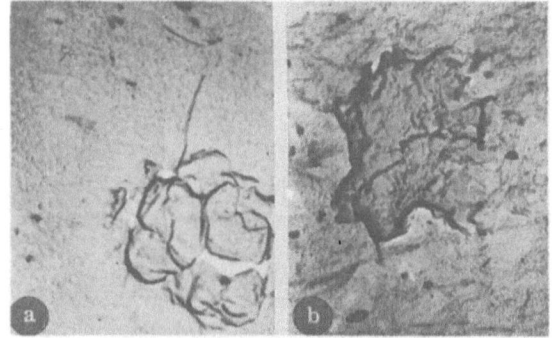

Fig. 7. Electron microphotograph of Ni—C alloy after annealing in air. Magnification × 40,000. Carbon replicas: a) 550° for 1 h; b) 650° for 1 h.

Discussion of Results

The change in the orientation of the spheroidal nuclei on passing from grain to grain indicates that the hexagonal form of the precipitating graphite crystals is determined by the structure of the metallic matrix in which crystallization takes place.

The form of the graphite crystals must be explained by some structural features of the nickel crystal lattice capable of causing the hexagonal form of the precipitating crystals. In face-centered metals there are in fact known defects (closed loops of

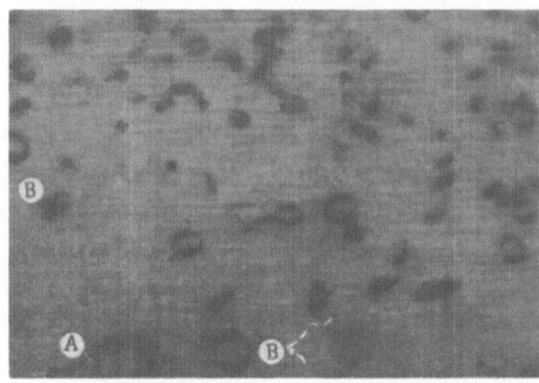

Fig. 8. Hexagonal loops of dislocations in aluminum [3].
Magnification × 44,000.

dislocations which have hexagonal shape and are disposed on the (111) planes (Fig. 8) [3]. It is also known that under certain conditions helicoidal dislocations [4] may be formed. It may be supposed that helicoidal, hexagonal dislocations form preferential precipitation centers for graphite. The parts of distorted lattice around a location may constitute sites at which carbon atoms easily separate. The crystal developing should have the form of a hexagonal toroid or helicoid, but in some cases, apparently, the inner aperture of the toroid may grow and form a crystal in the shape of a hexagonal prism.

The effect of hydrogen impurities on crystallization may be explained by assuming that hydrogen is absorbed in the crystal lattice of nickel at block boundaries, dislocation networks, and individual dislocations constituting channels along which carbon atoms are fed to the growing crystal and nickel atoms are removed. If these channels are blocked by absorbed hydrogen, the crystal nucleus should grow equally on all sides [6]. Thus when there is hydrogen present in the metal the uniform supply of carbon atoms to the growing crystal is ensured, and it grows simply by increasing in size. In the absence of hydrogen, carbon atoms are supplied to individual parts of the crystal surface, forming mounds, and as a result of these growths the regular form of the crystal is broken.

Conclusions

1. When a supersaturated Ni—C solid solution decomposes, graphite crystals precipitate in the form of spherulites if the alloy contains hydrogen, and otherwise as crystals of irregular shape.

2. The nuclei from which the spherulites develop are hexagonal prisms, pyramids, or tubes coiled into hexagonal toroids or helicoids. The orientation of the crystals is determined by the structure of the metallic matrix, and varies from grain to grain.

3. Possible centers for the formation of nuclei for spherulites are hexagonal loops of dislocations and helicoidal dislocations in the crystal lattice of nickel.

Literature Cited

1. H. Morrogh and W. J. Williams, J. Iron Steel Inst., 155:321 (1947).
2. A. Keller, J. Polym. Sci., 17:291 (1955).
3. P. B. Hirsch, J. Silcox, R. E. Smallman, and K. H. Westmacott, Phil. Mag., 3(32):897(1958).
4. S. Amelinckx, W. Botinck, W. Dekeyser, and F. Seitz, Phil. Mag., 2:355 (1957).
5. F. Seitz, Adv. Phys., 1(1):43 (1952).
6. K. P. Bunin, Yu. N. Taran, and A. V. Chernova, Cast Iron with Nodular Graphite (Izd. Akad Nauk Ukr. SSSr, 1955).

EFFECT OF THE NASCENT PHASE ON THE KINETICS
OF THE MARTENSITE TRANSFORMATION

É. I. Éstrin

The most characteristic trait of the martensite transformation, distinguishing it from other forms of phase transformation, is its markedly cooperative nature; the transformation from the old phase to the new takes place not by independent, individual displacements of separate atoms, but by a strictly ordered, regular, mutually connected transition, in the process of which the atoms are displaced from one another through distances not exceeding the interatomic value.

The cooperative character of the martensite transformation leads to two very important consequences. On the one hand, the martensite transformation (kinetics, structure, and hence properties of its products) is extremely sensitive to the state of the original phase and to actions affecting it, as was shown in a number of investigations into the effects on the martensite transformation of such actions as plastic deformation, phase hardening, and neutron irradiation [1]. The result of these actions is either the activation of the martensite transformations or a reduction in its intensity and stabilization of the original phase, depending on the nature of the action, its degree and temperature conditions, and peculiarities of the material. On the other hand, considerable distortion of the structure of the original phase is also found directly in the course of the actual transformation under the influence of its products [2–6, etc.].

Comparison of these data shows that an important role in the development of the martensite transformation must be played by the influence exerted on the original phase by the products of the actual transformation. In contrast to other phase transformations, where, in view of the high rate of relaxation processes, the effects of the newly developing phase may to a certain extent be neglected, in the martensite transformation this factor is very important, if not decisive, and many features of the martensite transformation may be associated with precisely this factor.

The present paper describes an experimental study of the effect of the newly developing phase on the progress of the martensite transformation.

Our examination of the effect of the newly developing phase on the progress of the martensite transformation was conducted as follows: The variation in the kinetics of the isothermal martensite transformation, under the influence of martensite previously accumulated under specified temperature conditions, was studied for a series of alloys of the N23G4 type (Table 1), having a martensite transformation of a sharply isothermal character, the austenite of which could be completely supercooled in the martensite range. We studied the role of such factors as the temperature and degree of previous transformation, the temperature of the subsequent ("control") transformation, the temperature and duration of intermediate annealing, etc.

The investigation was carried out on a recording magnetometric system specially developed for this purpose, by means of which the progress of the martensite transformation at various temperatures was followed [7].

In studying the effect of the martensite previously accumulated at the lowest temperature of the martensite range (-196°), on the subsequent development of an isothermal martensite transformation at a higher temperature (-90°), direct experimental data were obtained to show that the martensite transformation possessed a clearly autocatalytic character (Figs. 1 and 2); the presence of martensite which formed at -196° led to a sharp activation of the subsequent transformation at a higher temperature (Fig. 1); the initial rate of transformation (Fig. 2) rose extremely strongly, and the amount of martensite formed in the process of heating from -196 to

TABLE 1. Chemical Composition of Alloys Studied

Alloy	Chemical composition, %			
	C	Ni	Mn	Fe
1	0.03	23.6	3.6	Balance
2	0.05	23.6	3.5	
3	0.02	22.8	4.0	
4	0.02	22.7	4.3	

-90° and in the process of isothermal holding at -90° increased. The activation effect rapidly increased as the amount of martensite accumulated at -196° rose to 8 or 10%, and thereafter gradually fell.

The activating action of previously accumulated martensite appears most sharply in the initial stages of the isothermal process. With increasing duration of isothermal holding, the difference between the transformation rates in activated and nonactivated samples gradually falls.

The autocatalytic character of the martensite transformation also appears in the very course of the isothermal martensite transformation, which takes place at a rate increasing markedly with time in the first moments of isothermal holding.

The effect of the activation of the isothermal martensite transformation after partial transformation at a lower temperature was also established in Fe−Ni−Cr alloys of the N23Kh4 type; in these, just as in Fe−Ni−Mn alloys, the martensite transformation has an isothermal character and (for due choice of alloy composition) may be completely supercooled. The laws of activation in the Fe−Ni−Cr alloy were completely analogous to those holding for Fe−Ni−Mn alloys. These data indicate the general character of the activation phenomenon.

In studying the effect of the low-temperature (-196°) martensite transformation on the kinetics of subsequent isothermal transformations, it was shown that the presence of the products of the low-temperature transformation led to a sharp change in the kinetics of a subsequent martensite transformation (Fig. 3). The transformation interval in the higher-temperature region broadens substantially (more than 100°), and the initial rate of the isothermal transformation rises sharply.

The most substantial rise in the initial transformation rate (by two to three orders) is found at temperatures above the temperature of the maximum transformation rate in non-activated (high-annealed) austenite. At temperatures close to the lower limit of the martensite range, the effect of activation appears more weakly. The general character of the temperature-dependence of the transformation rate (curve with a maximum) is maintained after activation, but the maximum is displaced (by 30 to 50°) to the high-temperature side. The temperature-dependence of the effects of isothermal transformation for a finite length of time varies analogously.

An estimate of the effect of the degree of previous low-temperature (-196°) transformation on the position of the martensite point showed that the accumulation of even 1% martensite at -196° leads to a sharp rise in the martensite point. Further increase in the amount of preliminary (-196°) martensite to 10% led to only a slight additional rise in the

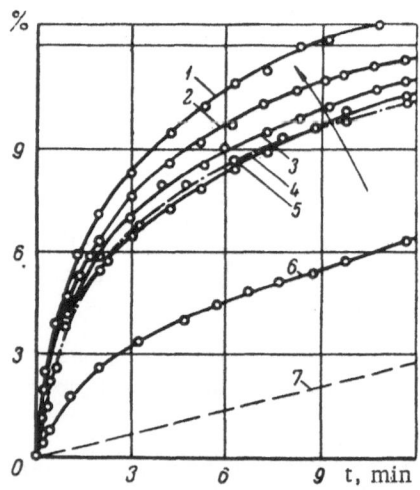

Fig. 1. Isothermal martensite transformation at -90° after partial transformation at -196°. Numbers on curves indicate amount of martensite accumulated at -196° (alloy 1): 1) 11%; 2) 9.3; 3) 5; 4) 16.8; 5) 3; 6) 1; 7) 0%.

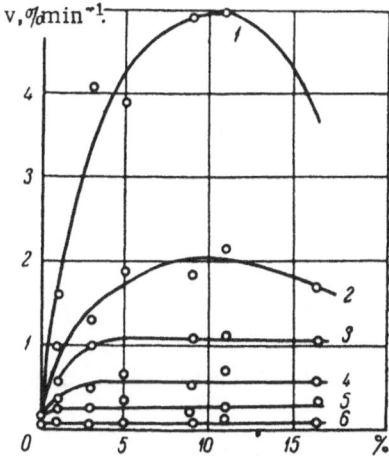

Fig. 2. Variation of the rate of isothermal transformation at -90° (in its various stages) with the amount of martensite previously accumulated at -196°. Numbers on the curves show the isothermal holding times for which the rate is determined (alloy 1): 1) 0 to 1 min; 2) 1 to 2 min; 3) 2 to 3 min; 4) 4 to 5 min; 5) 10 to 11 min; 6) 30 to 35 min.

martensite point. Further increase in the amount of preliminary (-196°) martensite to 10% led to only a slight additional rise in the martensite point. An increase of more than 50° in the martensite point after sharp cooling to -196° was also established in manganese steel 50G10 (0.51% C, 9.1% Mn, and 1.97% Cu, T_M = -87°).

As a result of studying prolonged isothermal transformation at a number of temperatures (-78, -100, -121°C) before and after activation, it was established that, despite a very sharp (many tens of times) rise in the initial rate of isothermal transformation after activation, the final effects of isothermal transformation in activated and nonactivated samples differed very little (Fig. 4).

In order to find to what extent activation reduced hysteresis in the martensite transformation (interval between the thermodynamic equilibrium temperature of the γ - and α-phases, T_0, and the martensite point T_M), the temperature T_A for the onset of the $\alpha \rightarrow \gamma$-transformation was determined, and the temperature T_0 of the alloys studied was deduced from the data on T_M and T_A (Table 2). As seen by comparing T_0, T_M, and T_M' (martensite point after activation), the activation leads to a considerable reduction in hysteresis (from 280 to 140°), but does not entirely remove it.

On determining the position of the upper boundary of the $\gamma \rightarrow \alpha$ transformation under the influence of deformation (M_d point) and comparing the mutual disposition of the M_d and T_M' points of

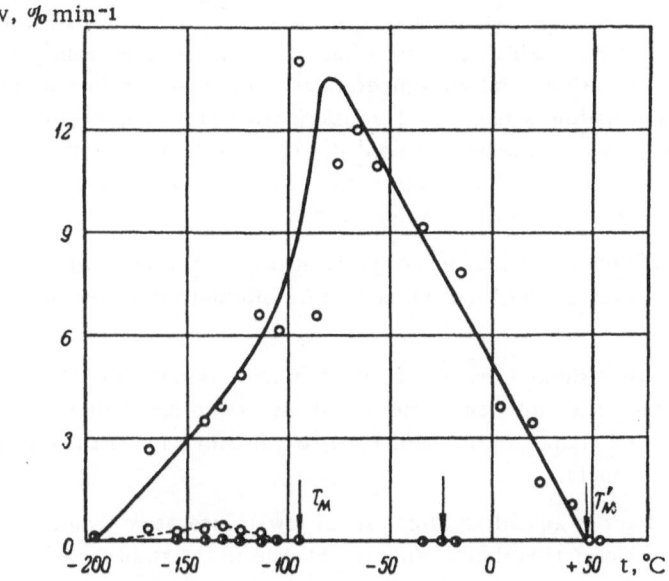

Fig. 3. Variation of the temperature-dependence of the initial rate of isothermal transformation under the influence of a partial transformation at -196°, $M_{-196°}$ = 4.3% (alloy 3). The maximum transformation rate before activation is shown by a broken line.

TABLE 2. Comparison between the T_M, T_M', T_0, and M_d Points (in °C) of Fe−Ni−Mn

Alloy	T_M	T_M'	T_A	$T_0 = \dfrac{T_M + T_A}{2}$	M_d	$T_0 - T_M$	$T_0 - T_M'$	$T_0 - M_d$
1	−80	55	470	195	35	275	140	160
2	−80	60	475	195	35	275	135	160
3	−90	50	490	200	25	290	150	175
4	−70	15	470	200	15	270	185	185

the alloys (Table 2), it was found that the M_d point in the alloys studied is close to the T_M' point and considerably (more than 150°) below the T_0 point.

Examination of the role of the temperature conditions of the preliminary transformation in the development of the activation effect showed that partial transformation at -110 and -150° (just as transformation at -196°) led to the activation of subsequent transformations at -78°, but as the temperature of the preliminary transformation rose the effect of activation gradually diminished.

A study of the effect of the martensite arising at a temperature close to the martensite point on subsequent development of the transformation at lower temperatures showed (Fig. 5) that the presence of the martensite accumulating at the "high" temperature (-78°) not only did not lead to activation of the subsequent transformation at lower temperatures (-110, -155°), but even retarded the process; the initial transformation rate at -110 and -155° rapidly fell as the amount of martensite previously accumulated at -78° rose, and the effect of isothermal transformation diminished. Thus we have direct experimental evidence that, quite apart from activation, the presence of martensite (even in very small quantities) may also retard a later martensite transformation.

Our study of the role of the relative temperatures of preliminary and subsequent transformations in causing activation or stabilization (Fig. 6) shows that it is not so much the absolute temperature of the preliminary transformation which decides which of these effects (activation or retardation) should prevail in the subsequent course of the process, as the relative temperatures of the two transformations; whatever the absolute temperature of the first transformation, a subsequent transformation at a higher temperature is activated and a subsequent transformation at a lower temperature retarded. The effect of activation or retardation is the stronger, the greater the temperature difference between the first and second transformations, the activation effect being more sharply expressed than that of retardation.

If the preliminary transformation occurs successively at several temperatures instead of at one, the resultant effect (activation or retardation of the subsequent transformation) is decided mainly by the last of such transformations.

Metallographic examination showed that the activation and retardation of the martensite transformation under the influence of previously accumulated martensite is associated not with a change in the dimensions of the martensite crystals but with a change in the number arising in unit time (i.e., with a change in the rate of formation of martensite crystal nuclei).

Detailed study of the effect of annealing after partial low-temperature transformation on the kinetics of a subsequent martensite transformation showed that the change in the stability of the austenite resulting from the annealing had an extremely complex nature and took place in several stages. The distortions responsible for the activation of the martensite transformation have a relatively high temperature resistance and are only fully removed on heating to 350 or 400°C. The activation energy of this process is ~30 kcal/ g·atom. Higher annealing (450° to 550°C) effected immediately after the forward martensite transformation or after a rapid reverse $\alpha \rightarrow \gamma$-transformation is accompanied by very strong austenite stabilization. The

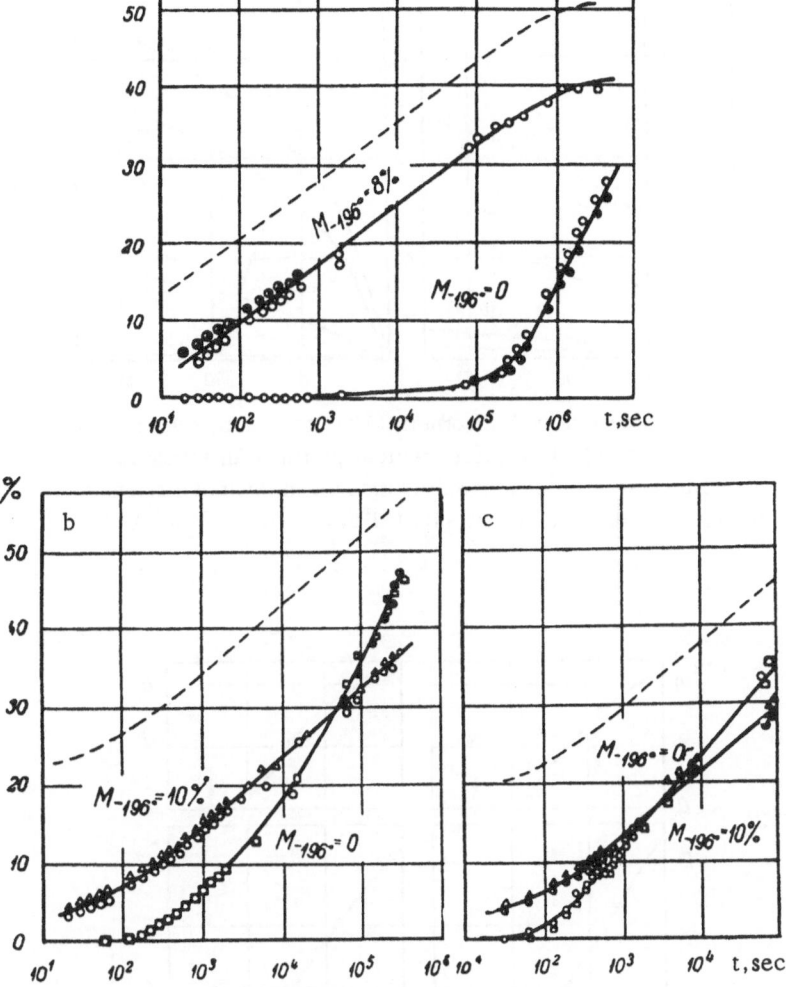

Fig. 4. Prolonged isothermal transformation [a) at -78°, b) -100°, c) -121°C] of activated and nonactivated austenite. Continuous lines represent the isothermal effects of transformation; broken lines represent the total amount of martensite in activated samples (alloy 1).

stability of the austenite after such annealing considerably exceeds that of austenite stabilized only by carrying out a cycle of foreard and reverse $\gamma \rightarrow \alpha \rightarrow \gamma$-transformations. For still higher annealing temperatures (650° and upwards), there is a gradual restoration of the ability of the austenite to undergo martensite transformation.

Thus our results indicate that the kinetics of the martensite transformation are substantially determined by the action which the products of the transformation exert on the original phase. This action may result both in a sharp activation of a later transformation (autocatalytic effect) and in its retardation (stabilization of the original phase), depending on the amount of the martensite phase and the conditions of its formation. Analysis of the laws obtained enables us to give the following treatment.

The activation of the martensite transformation after a partial transformation (autocatalytic effect) is apparently explained by the fact that the generation of subsequent martensite crystals is made much easier in the elastically distorted parts of the austenite adjacent to existing martensite crystals. This is to a certain

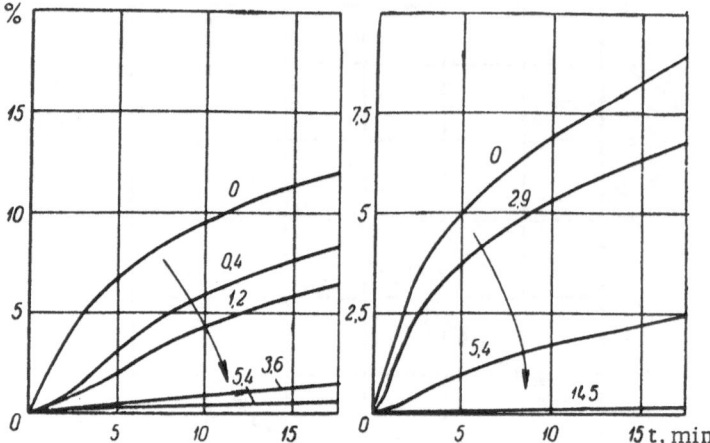

Fig. 5. Retardation of the isothermal martensite transformation
at a) -110° and b) -155° after previous partial transformation at
-78°. Numbers on the isotherms indicate the amount of martensite
previously accumulalated at -78° (alloy 1).

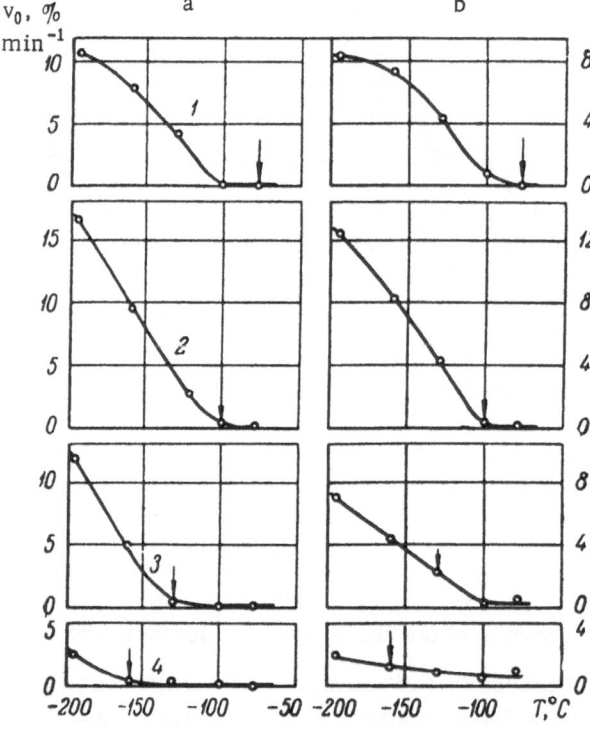

Fig. 6. Variation of a) the initial rate and b) the effects of isothermal
transformation at 1) -78, 2) -100, 3) -130, and 4) -160° with temperature of previous transformation. In all cases 3.5% martensite was
accumulated during the preliminary transformation (alloy 3).

extent analogous to crystallization on a substrate, but in the case of the martensite transformation the main part is played by the increase in elastic energy rather than surface energy as in crystallization. The higher the level of elastic distortions arising during the preliminary transformation, the more the subsequant transformation will be activated. This evidently explains the function of the relative temperatures of the preliminary and subsequent transformations in producing activation or stabilization: the lower the temperature of the preliminary transform..tion and the higher the corresponding yield point of the austenite,* the greater elastic distortions develop in the austenite during transformation, and the more the subsequent transformation is activated. On the other hand, when the preliminary transformation temperature is higher than that of the subsequent transformation, the elastic distortions developing in the austenite are of a lower order than those arising in the subsequent low-temperature transformation, and an apparent retardation of the control transformation results.

The increase in the martensite point following a low-temperature transformation, just as the course of the transformation above the martinsite point during plastic deformation, may be associated not only with purely kinetic causes but also with the fact that, in both plastic and low-temperature deformation, the austenite accumulates a considerable elastic energy, leading to a change in the thermodynamics of the transformation (increase in its thermodynamic stimulus). The mutual disposition of the martensite point T'_M of the activated austenite and the point M_d is evidently determined by the ratio of the energies stored in the austenite during the preliminary low-temperature transformation and in the process of plastic deformation.

Literature Cited

1. O. P. Maksimova, Probl. Metalloved. i Fiz. Met., (7):246 (1962).
2. Ya. M. Golovchiner and Yu. D. Tyapkin, Dokl. Akad. Nauk SSSR, 93(1):39 (1951).
3. B. Edmonson and T. Ko, Acta Met., 2(2):235 (1954).
4. A. I. Alfimov and A. P. Gulyaev, Izv. Akad. Nauk SSSR, Otd. Tekhn. Nauk, (4):93 (1954).
5. L. G. Khandros, Fiz. Met. i Metalloved., 3(1):479 (1955).
6. M. E. Blanter and P. V. Novichkov, Metalloved. i Obrabotka Metal., 6(6):11 (1957).
7. É.I. Éstrin, Zavodskaya Laboratoriya, 27(11):1423 (1961).

---

* Study of the temperature-dependence of mechanical properties in the γ-phase of Fe—Mn—Ni alloy showed that at low temperatures (room temperature downwards) there is a sharp rise in the yield point as temperature falls. In the temperature range studied, the temperature dependence of the yield point is satisfactorily described by the expression $\sigma_t = 100\ T^{-2/3}$, where T is the temperature in °K.

CHANGE IN MOSAIC STRUCTURE ASSOCIATED WITH THE DISTRIBUTION OF IMPURITIES IN GROWING SINGLE CRYSTALS AND BICRYSTALS FROM THE MELT

A. A. Kralina and V. O. Esin

Detailed study of the conditions of competition between two crystals with specific orientations along the direction of growth is usually carried out for bicrystals. It was shown by Chalmers [1] that, for slow rates of solidification, the boundary between the crystallites in a bicrystal remains parallel to the direction of growth, independently of the orientations of the crystals. With increasing growth rate, however, this boundary inclines toward one of the crystallites, namely, to that for which the preferential growth direction makes the larger angle with the direction of the temperature gradient. The deviation increases with increasing growth rate and increasing difference between the relative orientations of the crystallites with respect to the growth direction.

In order to explain the appearance of an inclined boundary, two fundamental mechanisms were proposed by Chalmers [2] and Tiller [3], respectively, for its formation. The mechanism proposed by Tiller is apparently the more correct. The main factor connecting these is the assumption that there must be a step between the crystallization fronts of neighboring crystallites in a bicrystal (Fig. 1), especially in the case of the crystallization of a melt containing impurities. The existence of this step was confirmed experimentally [4, 5]. The mechanism of step formation and the resultant appearance of an inclined boundary is as follows. When a crystal grows from a melt containing impurities, the layer of liquid adjacent to the surface of the crystallization front is enriched with impurities having a distribution coefficient smaller than unity. For a sufficiently high rate of growth, conditons may arise such that there appears in front of the crystal surface a layer of melt actually supercooled as a result of the nonuniform distribution of impurities caused by the finite rate of diffusion-equalization of the concentration inhomogeneity in the melt.

It was shown earlier [6—8] that, for critical conditions of growth, the smooth surface of the crystallization front will break up into a surface consisting of a network of cells projecting into the melt. This is due to the development of concentration supercooling and the consequent instability of the smooth surface. The cells

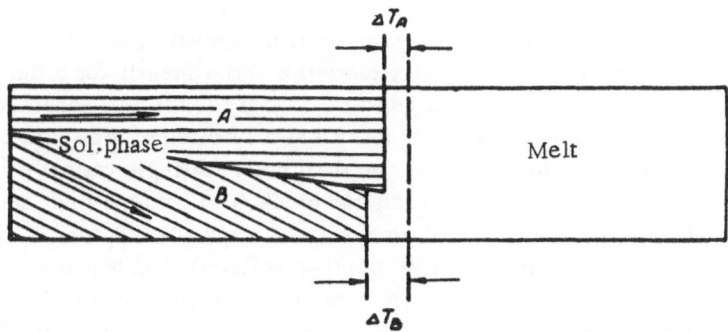

Fig. 1. Schematic representation of the growth of a bicrystal with an inclined boundary and the appearance of a surface step (ΔT) at the crystallization front.

repel the impurities to their boundaries, redistributing the impurity which has built up at the crystal surface and which is impeding its further growth. The small cross section of the cells ensures the repulsion of impurity to the boundaries in a relatively short time. Thus the concentration of impurity at the crystal-melt surface of separation diminishes, and hence so does the concentration supercooling. In fact, the more impurity there is concentrated at the boundaries, the less there is in the impurity-enriched layer of the melt, and hence the lower is the impurity concentration in front of the surface of the growing crystal. The concentration supercooling is affected by the conditions governing the repulsion of impurity to the boundaries of the cell substructure (in the final reckoning, the capacity of the boundaries) and the relative sizes of the cells on the surface of crystallites oriented in different ways (i.e., the total extent of the boundaries belonging to unit area of surface).

For face-centered cubic crystals, the most effective growing direction is the [100]. This direction is an axis of symmetry for four planes of layers lying along the [111], i.e., an axis relative to which the diffusion-redistribution of impurity occurring during layer growth is symmetrical. It follows that a point on the crystal surface with orientation [100] will have a more effective reduction in the impurity concentration than one on a surface with any other orientation [hkl].

The reduction in impurity concentration at the surface raises its solidification temperature. Hence the (100) surface can grow at a higher temperature than an (hkl) surface, and this leads to the appearance of a surface step during the growth of the bicrystal. Thus the presence of impurity, resulting in the appearance of a region of concentration-supercooling of the melt, means that the surface of separation of one crystal is stable for a higher temperature than that of a crystal with a less preferred orientation. We can explain the development of texture during the crystallization of a melt containing impurity by means of the surface step. From the example of the bicrystal with an inclined boundary, we see that one crystallite (with preferred orientation) in moving forward will creep up on the retarded crystallite and finally squeeze it out of the sample.

Effect of Growth Rate and Crystallite Orientation on the Perfection of the Crystallites

A change in the conditions of crystallization in front of the surface of the growing crystal, for example, the appearance of an inclined boundary in a bicrystal or a change in the orientation of a single crystal, cannot affect the degree of perfection of the crystals being grown from the melt. In general, the degree of perfection of crystals grown at different rates but otherwise under the same conditions is the greater, the lower the growth rate (i.e., the closer to equilibrium the conditions in which the crystal is grown). The solid phase developing on crystallization may appear either as a single crystal or a polycrystalline mass. In the latter case there will be considerably greater distortion of the crystal lattice per unit volume of solid phase (more intergrain and block boundaries) than in a single crystal. It is known that if we grow a single crystal, then the facility of forming it depends on the crystallographical orientation of the seed, or if the growth takes place without a seed (by random generation) then the probability of forming single crystals varies greatly for different orientations. This variation increases on passing to higher rates of growth. If we try to set up a schematic representation of the degree of defectiveness in the crystal lattice being formed as a function of growth rate, then this will appear roughly as follows. For low solidification rates, we shall obtain a single crystal with coarse blocks and low relative disorientation. On raising of the growth rate the blocks will become finer and their relative disorientation will increase. Then boundaries with still greater free energy will appear. This is the moment for the appearance of impurity substructure (first strip-like and then cellular), and hence an increase in the non-uniformity of impurity distribution in the crystal. At the same time the density of dislocations and vacancies rises inside the blocks themselves. For a still greater rate of forming the solid phase from the melt, dendritic solidification will develop, and finally solidification in the form of a single crystal will become altogether impossible. It will become possible for random crystals to be generated in the melt in front of the surface of the growing single crystal, and a polycrystalline mass will be formed. The solid phase developing in poly-crystalline form will have a still greater degree of defectiveness per unit volume, in the form of large-angle intergrain boundaries, than in the case of the single crystal. The moment of transition into polycrystalline form (which characterizes the instability of solidification as a single crystal) occurs at different solidification rates for single crystals with different crystallographic orientations along the growth direction. For single crystals with preferred orientation (for example, <100> for face-centered and body-centered cubic crystals), this transition occurs for considerably greater rates than for single crystals with any other crystallographic orientation <hkl>. This is evidently because the possibility of generating random crystals increases with increasing super-cooling (in the present case concentration supercooling). For a crystal with high indices of orientation, the growth rate for which the crystal can just eliminate all the concentration supercooling (by redistributing the impurities on the surface) is smaller than for a crystal with low indices of orientation (especially with a pre-ferred growth direction). Hence the degree of concentration supercooling in the melt in front of a crystal with high orientation indices will be larger for the same rate than that in front of a crystal with low orientation indices, and the probability of random generation will also be larger.

Even for slow rates, however, when single-crystal solidification of the melt is still possible, the degree of perfection of two crystals grown at the same rate but with different orientations must be different. This may well be imagined by considering the change in the conditions of crystallization in the formation of a bicrystal. As we know, in growing two crystals in a bicrystal in completely identical conditions of crystallization, begin-ning with a quite definite growth rate, we may expect different degrees of perfection in the resulting crystals. This will take place for the growth rate from which the boundary in the bicrystal deviates from the direction of growth in one of the crystallites. Actually the deviation of the boundary begins from the moment at which a surface step appears between the crystallites in the bicrystal, and hence from the moment at which a dif-ference arises between the degrees of concentration supercooling of the melt in front of the crystallization fronts of the two crystals. The formation of crystals from a melt supercooled to different degrees must considerably affect their degree of perfection. In the absence of such a surface step, i.e., when we have a parallel boundary between the crystals in the bicrystal, the relative degree of perfection of the developing crystals should not vary greatly over a wide range of growth rates. There is only one case [9] in the literature in which the dis-location density has been found experimentally as a function of growth rate for germanium single crystals with different crystallographic orientations in the direction of the temperature gradient (Fig. 2). We see from Fig. 2 that the smallest dislocation density belongs to a single crystal growing in the [112] direction, which constitutes the direction for the growth of dendrites in germanium.

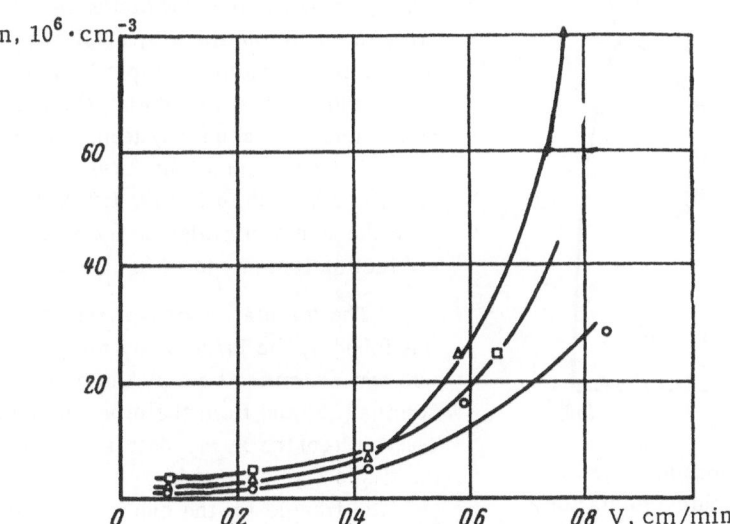

Fig. 2. Dislocation density as a function of the rate of growth of a germanium crystal for three crystallographic directions [9].

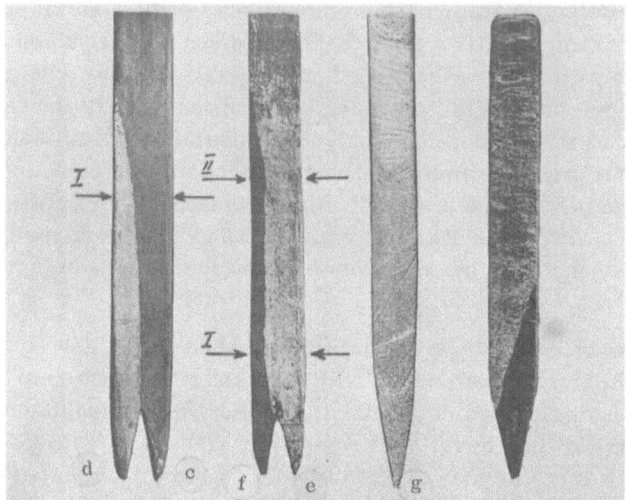

Fig. 3. Typical form of grown aluminum crystals. The
bicrystal c-d was obtained at a growth rate of 1.2 mm/
min, the bicrystal e-f at 0.35 mm/min, the single crystal
g at 0.8 mm/min, and the bicystal with the sharply
sloping boundary at 3.6 mm/min.

X-ray Investigation of the Effect of the Conditions of Crystallization on the Degree of Perfection of Aluminum Single Crystals and Bicrystals

As material for the investigation we took 99.95% pure aluminum. Single crystals and bicrystals were grown in aluminum oxide boats. The boats were prepared in two forms: with one and two sharpened ends.

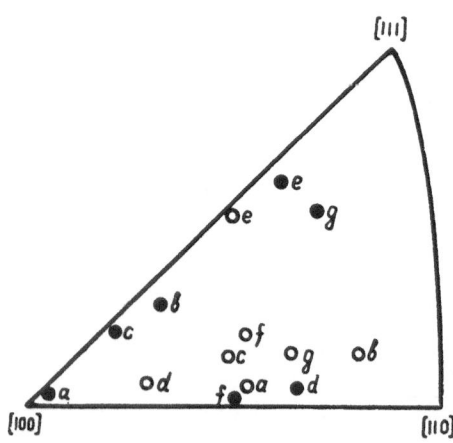

Fig. 4. Orientation of aluminum crystals
obtained: ● crystallographic orientation
in direction of growth; ○ orientation of
normal to free surface of crystals. Letters
on the points correspond to crystal notation
in Figs. 3 and 6.

Aluminum samples were cut in the shape of the boats with dimensions 125 by 11 by 3.5 mm for single crystals and 125 by 15 by 3.5 mm for bicrystals. The single crystals and bicrystals were obtained by displacing the furnace relative to a stationary molten sample lying inside the boat in a quartz tube with a sealed end; the other end of the tube was connected to a vacuum system. The crystals were grown in a 10^{-5} mm Hg vacuum at varying furnace-displacement rates: 0.1, 0.35, 0.5, 0.8, 1.2, 2.2, and 3.6 mm/min. This gave the single crystals and bicrystals with different boundary inclinations as shown in Fig. 3 and 6.

The orientation of the resultant crystals of both kinds was found by the Laue X-ray method, and the mosaic structure was examined both by the method of Osienko and Sosnina [12] and from the broadening of the Laue reflections. The results of the X-ray determination of the orientation of the single crystals and bicrystals are shown on the stereographic triangle for the cubic lattice of aluminum (Fig. 4). In determining the mosaic angle, we used the principle of the two-crystal spectrometer, the crystal under examination being disposed parallel to a calcite monochromator, rigidly connected to the X-ray tube. The total angular range of

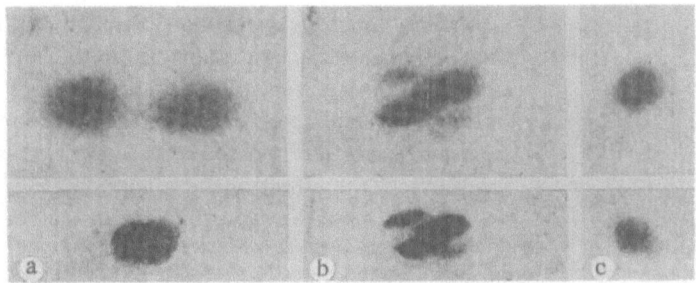

Fig. 5. Fine structure of Laue reflections obtained from adjacent crystallites of a bicrystal with a strongly inclined boundary. a and b) For cases of tapering-out crystallites in the bicrystal; c) From an adjacent crystallite in the same bicrystal with preferred orientation in the direction of growth (the lower and upper plates are distinguished by different Θ in the X-ray photographs). Magnification $\times$ 10.

reflection from the mosaic crystal in the region of the Bragg angle was determined from the number of reflections taken on a plate moving simultaneously with a rotation of the crystal through 10". The divergence of the primary beam was determined in an analogous way using a calcite standard. The maximum mosaic angle of the crystal was determined as the difference between the total angular range of reflections and the divergence of the incident monochromatic beam (λK_α Mo). It was noticed with good reproducibility that the block disorientation of the mosaic of single crystals was very different at the beginning and at the end of the sample. For example, for a single crystal grown with the furnace moving at 0.8 mm/min (see Fig. 3) and with the orientation indicated in Fig. 4, the disorientation increased more than three times from the beginning of the sample to the end.

Bicrystals also show an increase in relative imperfection from the beginning of the sample to the end, and the degree of imperfection of the adjacent crystals also depends on the inclination of the boundary.

In the bicrystals studied, a substructure of the cellular or ribbon type, i.e., a macromosaic, was observed visually. The macromosaic affected the X-ray photographs used in determining the total disorientation angle by the monochromatic method in such a way that the sequence of reflections suddenly cut off and then reappeared after a nonperiodic interval. This is evidently associated with the reflection of the primary beam first by one element of substructure, then by another, and so on. The effect of the macromosaic makes the determination of the total disorientation of the elements of substructure in the single crystal more difficult, since the break in the sequential disposition of the reflections may be taken to be the end of the angular scatter. In view of this, in addition to using the method described, we determined the degree of perfection as a function of crystallite orientation from the broadening of the spots on the Laue photographs. To a first approximation we may evidently suppose that the total disorientation is proportional to the area of the Laue reflection. In the investigation we chose reflections in which factors affecting the intensity and broadening, (θ) and (hkl), were the same. The photographs were taken using a comparatively fine X-ray beam (diameter of beam incident on the sample 0.5 mm). In order to be able to judge the change in the relative degree of perfection of adjacent crystallites from the broadening of the reflections, both crystallites were photographed in regions lying at the same distances from the beginning (or end) of the bicrystal. During the exposure, the sample was gradually moved relative to the primary beam, perpendicular to the direction of growth. Preliminary study of the change in the degree of perfection of individual crystals along the direction of growth did not show any spasmodic changes for distances of 6 to 8 mm. Hence the relative perfection perpendicular to the direction of growth may be determined at any point without danger of any unforeseen sharp change in the size of the reflections from point to point.

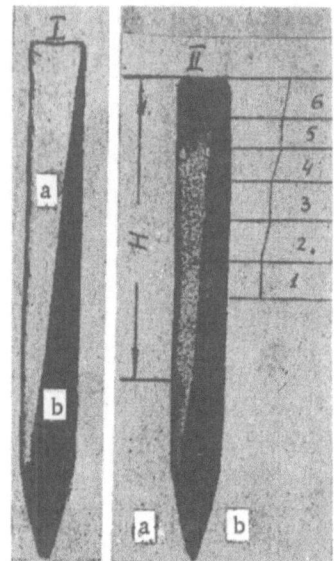

Fig. 6. Surface of bicrystals grown at different rates: I) Bicrystal a-b, growth rate 2.2 mm/min; II) Same bicrystal, but part H was melted and crystallized again at 0.5 mm/min; the boundary became undulating, in some part (1, 3, 5) being parallel to the direction of growth of the bicrystal, and in others (2, 4, 6) sloping into crystal b.

The sample was displaced through distances equal to or somewhat less than the beam diameter, so that the disorientation of both mosaic blocks and substructural elements was represented in the reflection. The exposure and total displacement (3 to 6 mm) during the photograph were the same for both crystals. In examinations by this method, the effect of the macromosaic is apparent not only in the broadening of the spots but also in their splitting. This is clearly visible from the variation of the spots in the Laue photographs of bicrystals shown in Fig. 5. This figure reproduces Laue reflections from adjacent crystallites of bicrystals with a strongly inclined boundary grown at the same rate. The lower reflections correspond to small θ and the upper to large θ (showing better resolution). For the case of crystals strongly deviating from the growth direction, the splitting of the reflection into two or four spots is clearly seen (Fig. 5). At the same time, reflections from a crystallite having preferential orientation in the growth direction show no marked changes (Fig. 5).

Figure 6, II shows bicrystal 1-b, with an inclined boundary, grown at a furnace-displacement rate of 2.2 mm/min. The degree of perfection of the adjacent crystals was determined in three positions. The results obtained by the first and second methods are given in Table 1. The maximum error in determining the degree of perfection was 5%. The two methods agree satisfactorily in showing that the relative degree of perfection in crystal b is worse than in crystal a, and the difference in the degree of perfection increases from the beginning of the crystal to the end. The considerable divergence in the values obtained in the first position was closer to the sharp end than in using the first method. The data from the determination of orientation (see Fig. 4) show that crystal a has preferred orientation in the growth direction while crystal b deviates from this.

In order to preserve the orientation but to obtain a less inclined boundary, part of the bicrystal a-b (H) was melted, and the new bicrystal shown in Fig. 6, II grown with a slower furnace displacement (0.5 mm/min). The new bicrystal has another boundary consisting of inclined and parallel parts, six in all. The variation in

TABLE 1. Relative Degree of Perfection of Adjacent Crystallites in Bicrystal a-b with Inclined Boundary (Growth Rate 2.2 mm/min) as Determined by Two Different Methods

Means of determining relative perfection of crystals	Relative position of point studied in the bicrystal		
	Beginning I	Middle II	End III
From broadening of angular range of reflection in the monochromatic method	1.5_0	1.6_5	1.7_5
From broadening of Laue reflections	1.2_5	1.6_4	1.7_5

TABLE 2. Relative Degree of Disorientation of Blocks in Neighboring Crystallites
of Various Bicrystals

Characteristic of bicrystal	Relative situation of point studied in the sample	$\langle hkl \rangle$	Area of Laue reflections in bicrystal X-Y		$\frac{S_y}{S_x} = \alpha$
			Crystal X	Crystal Y	
Bicrystal a-b with inclined boundary, growth rate 2.2 mm/min (see Fig. 6,I)	Beginning I	$\{113\}$	1.5_0	1.2_0	1.2_5
	Middle II	$\{113\}$	1.8_0	1.1_0	1.6_4
	End III	$\{113\}$	2.1_0	1.2_0	1.7_5
Same bicrystal a-b after partial recrystallization at 0.5 mm/min. Taken at various points of the new boundary with relatively parallel (1, 3, 5) and inclined (2, 4, 6) sections (see Fig. 6, II)	1	$\{113\}$	1.6_1	1.7_0	0.9_5
	2	$\{113\}$	2.6_6	2.6_5	1.6_1
	3	$\{113\}$	1.9_8	1.7_1	1.1_6
	4	$\{113\}$	2.9_6	1.7_4	1.7_0
	5	$\{113\}$	1.6_8	1.8_2	0.9_2
	6	$\{113\}$	1.5_0	1.2_2	1.2_3
Bicrystal c-d with inclined boundary and growth rate 1.2 mm/min (see Fig. 3)	End	$\{115\}$	1.2_5	0.8_0	1.5_6
		$\{204\}$	2.5_2	1.6_0	1.5_7
Bicrystal e-f taken in sections of the parallel boundary. Growth rate 0.35 sections mm/min (see Fig. 3)	Beginning I	$\{115\}$	1.0_8	1.1_0	0.9_8
		$\{313\}$	1.8_0	1.7_0	1.0_6
	End II	$\{115\}$	1.6_0	1.5_8	1.0_6
		$\{313\}$	2.8_6	2.7_3	1.0_4

the degree of perfection of these sections as indicated by the variation in the $\{113\}$ Laue reflections is shown in Table 2. The table shows that, in the sections touching along parallel boundaries (1, 3, 5) in crystals a and b, there is only a slight change in the degree of perfection, while in the sections touching along inclined boundaries the change is considerable. There is a certain diminution in the values of α in positions 5 and 6 as compared with the previous. This is because the total interval passed through by the beam in the crystal in these positions was reduced by a factor of almost two, owing to the reduction in the breadth of the tapering crystal. Hence in the last two cases the value of α fell because the relative proportion of the divergence of the actual beam in $\Delta\theta$ rose as the ratio of the total length of the region traversed to the beam diameter diminished. Nevertheless we see that the general tendency of the variation in the relative disorientation corresponding to the parallel and inclined boundaries is preserved. This tendency is also confirmed by results obtained for specially grown bicrystals with inclined and parallel boundaries, also given in Table 2.

The bicrystals with inclined (c-d) and parallel (e-f) boundaries were grown at furnace-displacement rates of 1.2 and 0.35 mm/min respectively (see Fig. 3). The data for bicrystal c-d indicate an increase in the degree of imperfection in the crystallite deviating more considerably from the direction of growth. In studying bicrystal e-f it was found that, for a small relative deviation of the orientations of neighboring crystallites in the bicrystal from the growth direction (i.e., for a parallel or nearly parallel boundary), there was either a slight change in the degree of perfection or none at all.

Discussion of Results

   Study of Single Crystals. The rise in the degree of block disorientation in a single crystal from its beginning to its end (part of the crystal last solidifying) is evidently connected with the existence of a natural segregation of impurities with a distribution coefficient smaller than unity, the concentration of which rises considerably in the last portions of the solidifying melt. This is the effect on which zone refining is based [10]. Hence different parts of the single crystal are formed from a melt with different impurity contents, and this, as we may suppose [11], must lead to a variation in their degree of perfection. In fact, on changing the super-cooling or amount of impurities in the melt, both the conditions and kinetics of growth and the very medium from which the crystal is being formed will change at the growing surface. For example, in the initial part of the sample, where the crystal originates, growth during the first moments takes place with a severely super-cooled melt. In the middle part of the sample, growth takes place mainly at the velocity of furnace motion, under conditions of less severe supercooling and with more or less constant impurity content. Finally, at the end of its growth, the crystal is formed from a melt considerably enriched by impurities repelled into it during the process of crystallization, and hence under conditions of large concentration supercooling. All this leads to a considerable variation in the degree of perfection of the single crystal along its length, as in fact observed experimentally.

   In connection with the question under consideration and the results obtained, it is interesting to analyze certain data of Ovsienko and Sosnina [12, 13] devoted to a detailed study of block structure in aluminum single crystals grown from the melt. These workers grew ten single crystals for each growth rate and determined the mosaic angles. As the authors themselves noted [12], for samples grown at the same rate the mosaic angles varied a great deal from sample to sample. Moreover, experiments showed that the data obtained for different parts of the same sample differed by 10% or more. It is especially interesting that for growth rates of 0.1 and 0.5 mm/min there was a great scatter in the values, while for growth rate 6 mm/min the results measured for different samples differed only slightly. For example, the maximum difference in the mosaic angles for single crystals grown at 0.1 mm/min was 90% of the mean values for each set; for single crystals grown at 0.5 mm/min the difference was 78%, and for 6.0 mm/min 27%.

   As it appears to us, this must be correlated with the probability of obtaining one orientation or another in the growth direction when growing single crystals with random generation. For very "drastic" conditions of crystallization (for example, at very high rates), as we know, only crystals with preferred orientations along the growth direction survive and grow. This leads, for example, to the development of texture in the columnar zone of an ingot and to a completely definite crystallographical orientation of the axes of the dendrites. With falling growth rate, the range of possible orientations of the single crystals conducive to successful growth widens. For very slow rates we may be able to grow single crystals starting with almost any orientation in the growth direction. The important feature here is the crystallographic orientation in the growth direction (in the direction of the temperature gradient), and any other definitions of preferred orientations (for example, the variation with growth rate of the indices of the crystallographic plane parallel to the surface of the single crystal being formed [14]) we can contribute little to the understanding of the problem. Hence for slow growth rates (0.1 and 0.5 mm/min) there is a considerable scatter in the perfection of the single crystals obtained; crystals starting in a direction close to the preferred direction of growth reach greater perfection than crystals grown under the same conditions but with less favorable orientation. For high growth rates (6 mm/min), the single crystals obtained have mainly orientations close to the direction of preferential growth, and this gives a considerably lower scatter in their degree of perfection.

   Study of Bicrystals. Study of bicrystals with parallel boundaries (when there is no boundary step on the surface of the crystallization front) shows that no marked difference is found in the degree of block disorienta-tion in neighboring crystallites having different crystallographic orientation in the growth direction. We may thus suppose that, although the kinetics involved in the building-up of different faces of the growing crystal do indeed differ, yet, under conditions of slow growth (close to equilibrium) from melts with the same degree of supercooling and impurity content, these have little effect on the relative degree of perfection of the crystal lattice being formed. On passing to high growth rates, or with increasing relative disorientation, a surface step forms between the two crystallites in the bicrystal on the crystallization front, and this results in the appearance of an inclined boundary between the crystallites. The existence of this step on the surface of the crystallization front means that the lagging crystallite grows from the melt with a larger impurity content and considerably greater supercooling than the leading crystallite. The result is that the block size in the crystallite tapering off in the bicrystal must be smaller than that of its more favorably oriented neighbor, and the block disorientation considerably greater. This difference in the degree of perfection of the solid phase being formed is directly connected with differences in the structure of the actual melt from which crystallite-formation is taking place.

At a certain stage in the crystallization of the bicrystal, the considerable accumulation of impurity in front of the less favorably oriented growing crystal leads to the appearance of impurity substructure. This appears in the X-ray photographs as a splitting of the Laue reflections into two, three, or four spots, depending on the number of substructural elements falling within range of the primary X-ray beam incident on the crystal surface. Thus the impurity substructure forms a kind of macromosaic consisting of single crystals, and these elements in turn have a definite degree of block structure. It is interesting to note that the degree of block disorientation inside each element of impurity substructure is usually less than or the same as that in the neighboring crystallite not having impurity substructure. This is evidently because the break-up of the smooth surface of the crystallization front and the formation of impurity substructure sharply reduces the concentration super-cooling, and the p ocess of crystal formation (the impurities being repelled to the newly forming boundaries of the substructure) takes place under more equilibrium conditions. Hence each element of mosaic of the less perfect crystallite in the bicrystal has approximately the same degree of perfection as the more perfect neighboring crystallite. In the previous stage of crystallization, however, when as yet impurity-substructure boundaries have not been formed, increasing the impurity in the melt at the surface of the growing crystallite leads to an increase in its degree of block disorientation (and hence a fall in block size) as compared with the more favorably oriented crystallite; this appears as a considerable broadening of the Laue reflections in the X-ray photographs.

We may conclude from this investigation that increasing the growth rate, or, for the same rate, increasing the relative deviation in the orientation of neighboring crystallites in the bicrystal from the direction of growth (i.e., increasing the inclination of the boundary), leads to an increase in the relative degree of imperfection of the crystallites (including both block and mosaic structures). A fundamental result worthy of mention is the importance of considering the crystallographic orientation in the growth direction in connection with the degree of perfection of a single crystal.

Literature Cited

1. B. Chalmer, Can. J. Phys., 31:132 (1953).
2. B. Chalmers, J. Metals, 6 (5):519 (1954); Trans. AIME, 200: 520 (1954).
3. W. A. Tiller, J. Metals, 9 (7, II): 847 (1957).
4. D. Turnbull, Trans. ASM, 42:282 (1950).
5. A. Rosenberg and W. A. Tiller, Acta Metal., 5 (7):565 (1957).
6. W. A. Tiller, K. A. Jackson, J. N. Rutter, and B. Chalmers, Acta Met., 1:428 (1953).
7. J. N. Rutter and B. Chalmers, Can. J. Phys., 31:15 (1953).
8. W. A. Tiller and J. W. Rutter, Can J. Phys., 34:96 (1956).
9. A. Kurtz, S. Kulin, and B. Averbach, J. Appl. Phys., 27:1287 (1956).
10. V. G. Pfann, Zone Melting [Russian translation], (GNTILChTsM, Moscow, 1960).
11. E. I. Sosnina and D. E. Ovsienko, Fiz. Metal. i Metalloved., (3):527 (1956).
12. D. E. Ovsienko and E. I. Sosnina, Fiz. Metal. i Metalloved., (3):374 (1956).
13. D. E. Ovsienko and E. I. Sosnina, Fiz. Metal. i Metalloved., (6):433 (1958).
14. D. E. Ovsienko and E. I. Sosnina, Fiz. Metal. i Metalloved., (2):270 (1956).

NATURE OF CERTAIN FEATURES IN THE MARTENSITE TRANSFORMATION

É. I. Éstrin

One of the most characteristic features of the martensite transformation, distinguishing this from other phase transformations, is the extinguishment of the transformation in isothermal conditions in the presence of the original phase, and the extension of the transformation over a range of temperature [1].

A number of hypotheses have been put forward to explain this property; despite their diversity, these can be divided into two groups.

According to one proposition, the reason for the extinguishment of the transformation at constant temperature is the change in the kinetic conditions of its development (increase in the potential barrier for the generation of martensite crystals, the exhaustion of "prepared" nuclei suitable for growth at the given temperature, and so forth). Here it is supposed that, in the presence of a sufficient number of nuclei suitable for growth, the transformation would proceed until all the original phase had been exhausted. This point of view subsequently developed into the so-called theories of heterogeneous generation [2-4], the theory of the "reaction path" [5, 6], the theories proposed by Knapp and Dehlinger in [7] and by Dehlinger in [8-10], the "operational" theory of Cohen [11, 12], etc.

According to the other point of view, the extinguishment of the transformation is connected with a change in the thermodynamic conditions of the system during the development of the transformation, in such a way that further development is thermodynamically unfavorable. An example of this is the concept according to which the transformation ceases because of a uniform pressure exerted on the original phase by the harder martensite phase, which has a larger specific volume [13—16]. It should be noted that this treatment of the martensite-transformation characteristic under consideration is not exhaustive, since it gives no explanation for the existence of the same transformation features in alloys where there is no accompanying increase in volume.

The question as to which of the points of view mentioned corresponds more to reality remains open. A number of facts, however, suggest that the extinguishment of the transformation in isothermal conditions is more of a thermodynamic than a kinetic phenomenon. These facts are: the existence of the effect in materials where the transformation occurs at a tremendous rate ("athermally") and is therefore not limited by kinetic factors; the almost-constant end effects of the isothermal transformation after reactions leading to an extremely sharp change (by several orders) in the initial transformation rate, i.e., in its kinetics, such as takes place, for example, after a partial low-temperature transformation [17]; the autocatalytic character of the martensite transformation [17,18], indicating the development, during the transformation, of a large number of sites favorable to the generation of martensite crystals.

In this connection, it is interesting to consider whether one can explain the characteristic features of the martensite transformation starting from purely thermodynamic considerations and taking account of the special properties of this kind of transformation.

Let us consider the energy balance of a system in which a martensite transformation is taking place:

$$\Delta F = \Delta F_0 + E_{dist} = -\Delta\mu M + e_{dist}M, \tag{1}$$

165

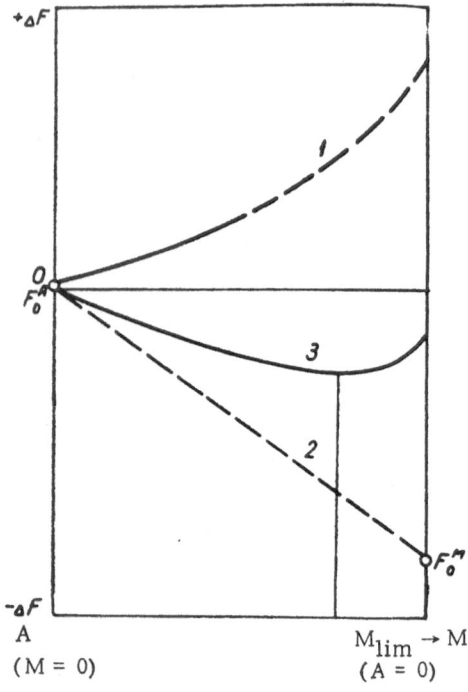

Fig. 1. Variation of 1) the energy of the distortions E_{dist}, 2) the change in the "chemical" free energy ΔF_0, and 3) the change in the total free energy ΔF of the system with the amount of martensite phase.

where ΔF is the total change in the free energy of the system in the course of the transformation, $\Delta F_0 = -\Delta\mu M$ is the change in free energy on account of the phase of volume M of the original phase into the phase with lower chemical potential μ, E_{dist} and e_{dist} are respectively the total and specific energies of the distortions arising in the system on formation of the martensite phase. The E_{dist} and e_{dist} include the surface and elastic energies and the increment in free energy associated with the development of all forms of structural defects in the system.

A characteristic of the martensite transformation from the thermodynamic point of view is the development and accumulation of considerable crystal-lattice distortions in the system in the course of transformation (the first as a result of the cooperative character of the transformation and its occurrence in a medium with high elastic properties, the second as a result of the development of the transformation at low temperatures when the rate of the relaxation processes is small), as shown directly in a number of experimental investigations [19−28, etc.]. Owing to the continuous accumulations of distortions in the system, there must be a continuous growth in the specific energy e_{dist} of the distortions as the transformation develops. Accordingly, in the expression (1) for the energy balance of the system, the second term, which should be written in the form $\int_{0}^{M} e_{dist}(M)\,dM$, increases with increasing amount of martensite phase more rapidly than the linear (with respect to M) term $\Delta\mu M$. Consequently for some value of M, for example (assuming that $e_{dist} = e_0 + kM^n$), for

$$M_{\lim} = \left(\frac{\Delta\mu - e_0}{k}\right)^{1/n}, \qquad (2)$$

where e_0 is the specific energy of the distortions at the onset of transformation (for M = 0), k and n (k>0, n<0) parameters defining the nature of the dependence of the specific energy of the distortions on the amount of martensite being formed, the free energy of the system expressed as a function of the amount of martensite phase has a minimum (Fig. 1), and the effective thermodynamic motive force of the process

$$\frac{d\Delta F}{dM} = -\Delta\mu + e_0 + kM^n \qquad (3)$$

vanishes. Further development of the transformation is thermodynamically unfavorable, and the transformation ceases. Other conditions being equal, the transformation rate dM/dt is proportional to the effective thermodynamic motive force of the process $d\Delta F/dM$:

$$\frac{dM}{dt} = -K\frac{d\Delta F}{dM} = -K(-\Delta\mu + e_0 + kM^n), \qquad (4)$$

where K is the kinetic coefficient. As the temperature is lowered, the motive force may begin to fall as a result of the rapid growth in the distortion energy. This may be one cause of the fall in the initial rate of the isothermal martensite transformation in the low-temperature range. In principle, the solution of equation (4) with respect to M gives the equation of the transformation isotherm.

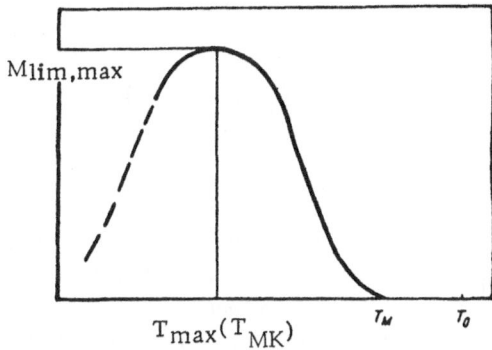

Fig. 2. Temperature-dependence of M_{lim}.

Analogous consideration of the growth of an individual martensite crystal enables us to explain the cessation of crystal growth on reaching certain sizes as well as the existence of thermelastic martensite. In order to resume the transformation, one must increase the thermodynamic motive force, i.e., increase the supercooling (lower the temperature).

From relation (2) we may obtain an expression describing the temperature-dependence of the limiting effects of the isothermal transformation, and also the martensite curve, for which we must find the temperature dependence of the quantities coming into the expression for M_{lim}. Here we must bear in mind that the specific energy of the distortions depends on the transformation temperature and increases as this falls (owing, for example, to an increase in the mechanical properties of the original phase, especially the yield point, and the fuller retention of the distortions arising). For an illustrative example we may assume, without limiting the generality, that e_0 varies with temperature on the same law as the yield point of the γ-phase, i.e., $e_0 = aT^{-m}$, where $m > 0$ [17]. Then

$$M_{lim}(T) = \left(\frac{L \ln \dfrac{T_0}{T} - aT^{-m}}{k} \right)^{1/n} , \qquad (5)$$

where L is the latent heat of the $\gamma \rightarrow \alpha$ transformation, and T_0 is the temperature for equilibrium of the α- and γ-phases.

Analysis of equation (5) shows that, first, M_{lim} vanishes at the point

$$L \ln T + aT^{-m} = L \ln T_0, \qquad (6)$$

clearly determining the martensite point T_M, and, secondly, M_{lim} does not vary monotonically with temperature, but has a maximum (Fig. 2). The latter circumstance is in accordance with the laws established empirically for alloys with an isothermal character of the transformation [29, 30]. It is clear that, in the case of a transformation taking place during continuous cooling, only the rising part of this curve can be realized (from the martensite point T_M to the maximum $T_{max} = (am/L)^{1/m}$), which also explains the origin and form of the martensite curve. In this case T_{max} corresponds to the temperature at the end of transformation with continuous cooling, T_{Mk}.

The maximum amount of martensite which can arise in a given material may be obtained by substituting for T_{Max} in (5):

$$[M_{lim}(T)]_{max} = \left[\frac{L}{k} \ln T_0 - \frac{L}{km} \left(\ln \frac{am}{L} + 1 \right) \right]^{1/n} . \qquad (7)$$

It follows from (7) that M_{lim} falls continuously with T_0, and for some value of T_0 determined by the condition

$$\ln T_0 = \frac{1}{m}\left(\ln \frac{am}{L} + 1\right)$$

reaches zero. The possibility of a transformation in this case is completely excluded. The conclusion regarding the fall in the limiting amount of martensite phase as T_0 diminishes is in accordance with the well-known law according to which, in materials with a low T_0 (and correspondingly low martensite point T_M) the transformation can never go to completion, and some of the high-temperature phase always remains, the amount of this being the larger, the lower the T_0 and T_M points for the given materials [31-32]. In contradistinction to the transformation on cooling, the martensite transformation of heating always proceeds until the original low-temperature phase has been completely used up. This is no doubt connected with the fact that, as the temperature rises, the specific distortion energy falls and can no longer retard the development of the process.

Thus an analysis of the thermodynamics of the martensite transformation, performed with due allowance for the characteristics of its mechanism and conditions of occurrence (elastic medium, low temperature), shows that the thermodynamic approach suffices to explain a number of fundamental features in the transformation: the extinguishment of the transformation in isothermal conditions, the characteristic temperature-dependence of the limiting effect of the isothermal transformation, the existence and form of the martensite curve, the beginning and end points of the transformation, and others.

Literature Cited

1. G. V. Kurdyumov, Quenching and Tempering Phenomena of Steel (Metallurgizdat, 1960).
2. J. C. Fisher, "Thermodynamics in Physical Metallurgy," J. ASM, p. 201 (1949).
3. J. C. Fisher, J. H. Hollomon, and D. Turnbull, J. Metals, 1 (10):691 (1949).
4. J. H. Hollomon and D. J. Turnbull, Advances in the Physics of Metals [Russian translation], 1:304 (1956).
5. E. S. Machlin and M. Cohen, J. Metals, (5):489 (1952).
6. M. Cohen, E. S. Machlin, and V. G. Paranjpe, Thermodynamics in Physical Metallurgy, ASM, p. 242 (1949).
7. H. Knapp and V. Dehlinger, Acta Met., 4, (3):289 (1956).
8. V. Dehlinger, Z. Physik, 149 (5):647 (1957).
9. V. Dehlinger, Physical Chemistry of Metallic Solutions and Intermetallic Compounds, Vol. 2 (London, HMSO, 1959), 4 of 1.
10. V. Dehlinger, Z. Metallkunde, 51 (6):353 (1960).
11. M. Cohen, Trans. Met. Soc. AIME, 212 (2):171 (1958).
12. L. Kaufman and M. Cohen, Uspekh. Fiz. Meta., 4:192 (1961).
13. C. Benedicks, J. Iron Steel Inst., 77 (2):233 (1908).
14. S. S. Shteinberg, Tr. Ural'sk. Fil. Akad. Nauk, (9):5 (1937) (editor's introduction).
15. E. Houdremont and O. Krisement, Arch. Eisenhüttenw., 24 (1-2): 53 (1953).
16. V. N. Gridnev and V. I. Trefilov, Collection of Scientific Papers of the Institute of Metallophysics (Academy of Science, Ukr. SSR), (8):29 (1957).
17. É. I. Éstrin, present collection, pp. 407-416.
18. O. P. Maksimova and É. I. Éstrin, Fiz. Metal. i Metalloved., 9 (3):426 (1960).
19. G. Wasserman, Arch. Eisenhüttenw., 6 (8):347 (1932/33).
20. Z. Nishiyama, Sci. Rep. Tohok. Univ., 23:637 (1934).
21. Ya. M. Golovchiner and Yu. D. Tyapkin, Dokl. Akad. Nauk SSSR, 93 (1):39 (1951).
22. L. S. Moroz, Hardening of Carbon-Free Alloys of Iron in Phase Transformations (Metallurgizdat, 1951).
23. B. Edmonson and T. Ko, Acta. Met., 2, (2):235 (1959).
24. A. N. Alfimov and A. P. Gulyaev, Izv. Akad. Nauk SSSR, Otd. Tekn. Nauk, (4):93 (1954).

25. L. G. Khandros, Fiz. Metal. i Metalloved., 1 (3): 479 (1955).
26. O. P. Maksimova and A. I. Nikonorova, Probl. Metalloved. i Fiz. Metal., 4:123 (1955).
27. L. S. Moroz, Fine Structure and Strength of Steel (Metallurgizdat, 1957).
28. M. E. Blanter and P. V. Novichkov, Metalloved. i Obrabotka Metal., 6:11 (1957).
29. G. V. Kurdyumov and O. P. Maksimova, Dokl. Akad. Nauk SSSR, 61(1):83 (1948).
30. O. P. Maksimova, E. G. Ponyatovskii, N. S. Rysina, and L. G. Orlov, Probl. Metalloved. Fiz. Metal., No. 5:56 (1958).
31. V. G. Vorob'ev. Heat Treatment of Steel at Temperatures Below Zero (Oborongiz, 1954).
32. A. P. Gulyaev, Heat Treatment of Steel (Mashgiz, 1960).